QUANTUM GODS

QUANTUM GODS

Creation, Chaos, and the Search for Cosmic Consciousness

VICTOR J. STENGER

Prometheus Books
59 John Glenn Drive
Amherst, New York 14228–2119

Published 2009 by Prometheus Books

Inquiries should be addressed to
Prometheus Books
59 John Glenn Drive
Amherst, New York 14228–2119
VOICE: 716–691–0133, ext. 210
FAX: 716–691–0137
WWW.PROMETHEUSBOOKS.COM

13 12 11 10 09   5 4 3 2 1

Library of Congress Cataloging-in-Publication Data

Stenger, Victor J., 1935–.
Quantum gods : creation, chaos, and the search for cosmic consciousness / Victor J. Stenger.
p. cm.
Includes bibliographical references and index.
ISBN 978–1–59102–713–3 (cloth : alk. paper)
1. Quantum theory—Religious aspects. 2. Consciousness—Religious aspects. I. Title.

BL265.P4S74 2009
215—dc22

2008054567

Printed in the United States of America on acid-free paper

CONTENTS

QUANTUM FLAPDOODLE AND OTHER FLUMMERY

Foreword by Michael Shermer

In the spring of 2004, while on a book tour that took me to the square block–sized Powell's Bookstore in Portland, Oregon, I appeared on KATU TV's *AM Northwest.* In the green room, where they mix authors with chefs, pet trainers, and dating self-help gurus with a not-so-healthy dose of junk food and coffee, I was introduced to the producers of a documentary film improbably named *What the Bleep Do We Know!?* They were pleasant enough fellows who seemed pleased to meet an editor of *Scientific American* because, they said, their film was about quantum physics. At the time I recall thinking, "A documentary film on quantum physics screening in a large public theater in competition with Hollywood films? This won't make it to the second weekend."

How wrong I was. *What the Bleep Do We Know!?* went on to become one of the highest grossing documentary films of all time. How can this be? Is the public suddenly interested in quantum physics? No. The explanation is to be found in the fact that the film is not really about quantum physics. The documentary's central motif is that we create our own reality through will, thought, and consciousness, which, according to the "experts" who appear as talking heads throughout the film (most of whom are not scientists, let alone quantum physicists), depends on quantum mechanics, that branch of physics so befuddling even to those who do it for a living that it can be invoked whenever something supernatural or paranormal is desired.

The Caltech Nobel laureate Murray Gell-Mann once described the misuse and abuse of quantum physics as "quantum flapdoodle." Examples from the film abound. The University of Oregon quantum physicist Amit Goswami, for example, says: "The material world around us is nothing but possible movements of consciousness. I am choosing moment by moment my experience. Heisenberg said atoms are not things, only tendencies." In my monthly column in *Scientific American*, I publicly challenged Dr. Goswami to leap out of a twenty-story building and consciously choose the experience of passing safely through the ground's tendencies. To my knowledge he has not taken me up on this experimental protocol. (In such an experiment you would most definitely want to be in the no-jump control group.)

Quantum flapdoodle infuses New Age gurus such as Deepak Chopra (whose quantum theory that aging is all in the mind is belied by the fact that he appears to be aging like everyone else), J. Z. Knight (masquerading as "Ramtha," the 35,000-year-old spirit warrior dishing out spiritual advice . . . for a price, of course), and Masura Emoto, the Japanese author of *The Message of Water*, who believes that thoughts change the structure of ice crystals—beautiful crystals form in a glass of water with the word "love" taped to it, whereas playing Elvis's "Heartbreak Hotel" causes a crystal to split into two. We are still awaiting replication of this research, not to mention publication of it in a peer-reviewed scientific journal. In the meantime, domo arigato, Mr. Emoto.

Moving beyond such New Age nuttiness, there have been serious attempts to link the weirdness of the quantum world (such as Heisenberg's uncertainty principle, which states that the more precisely you know a particle's position, the less precisely you know its speed, and vice versa) to mysteries of the macroworld (such as consciousness). A leading candidate to link the two comes from physicist Roger Penrose and physician Stuart Hameroff, whose theory of quantum consciousness is also featured in *What the Bleep Do We Know!?* According to this highly speculative conjecture, inside our neurons are tiny hollow microtubules that may initiate a wave function collapse that leads to the quantum coherence of atoms, causing neurotransmitters to be released into the synapses between neurons and thus triggering them to fire in a uniform pattern, thereby creating thought and consciousness. Since a wave function collapse can only come about when an atom is "observed" (i.e., affected in any way by something else), the idea is that the "mind" (either in your head or out there in space-time somewhere) may be the observer in a recursive loop from atoms to molecules to neurons to thought to consciousness to mind to atoms.... Maybe the entire universe is one giant mind that brings itself into existence by thought alone. Or maybe not.

When I first looked into this idea for my column on *What the Bleep Do We Know!?* in *Scientific American*, I called on the one man who knows both the science and the pseudoscience behind quantum physics, and that is Victor Stenger, who has single-handedly taken it upon himself to address each and every claim of the quantum flapdoodlists. For the Penrose-Hameroff conjecture, for example, Stenger explained to me that the gap between subatomic quantum effects and large-scale macrosystems is too large to bridge. Specifically, Stenger noted that for something to be described quantum mechanically, the system's typical mass m, speed v, and distance d must be on the order of Planck's constant h. "If mvd is much greater than h, then the system probably can be treated classically." Stenger computes that the mass of neural transmitter molecules, and their speed across the distance of the synapse, are about three orders of magnitude too large for quantum effects to be influential. There is no micro-macro connection. Q.E.D.

With this level of scientific and semantic precision Victor Stenger has taken on the God question in his book *God: The Failed Hypothesis*, addressing specific claims made for the Judeo-Christian God Yahweh in a systematic deconstruction that left them in tatters. As a result, Stenger's book rode the "New Atheist" Dawkins-driven wave to *New York Times* best-sellerdom. But Stenger added something new to this age-old debate, and that was a compelling argument for why there almost certainly is no God because of the contradictions inherent in the nature of God (or at least how God is portrayed in the Judeo-Christian worldview), as well as positive evidence that the universe does not need a creator-God.

But that was just a start for the polymathic Stenger. After all, Yahweh is just one among a pantheon of gods that have been conceived of in Western history, and in this, his latest masterpiece, Stenger picks up where he left off, addressing claims made for other gods, including and especially the sorts of arguments presented in *What the Bleep Do We Know!?* to which he devotes an entire chapter that also serves as a brilliant tutorial in quantum physics, of which the great Nobel laureate Richard Feynman once said that no one really understands. That may be, but Victor Stenger does as good a job as anyone ever has in explaining it, and in the context of why quantum physics—along with chaos theory, complexity theory, emergence theory, and other assorted branches of physics, biology, and neuroscience—does not get you to God.

There's more still in *Quantum Gods*. An important new development in theism is the use of quantum mechanics, the uncertainty principle, chaos, complexity, and emergence to make the case for how God acts in the world. That is, most theists do not believe in some airy fairy deity who lives in the hinterlands of the cosmos and never bothers with our trivial lives here on Earth; indeed, one of the most common arguments given for belief in God is divine providence—God reaches into our world from outside of space and time and interacts with us by performing miracles, answering prayers, and directing the flow of history, for example, bringing about the end of the world through an inexorable unfolding of events. How does God do this? Curious minds want to know. In the Age of Science, it's not enough to just say, "God works in mysterious ways." Serious theologians need to answer the scientist's question: how does God act in the world? For example, a scientist would want to know how God cures cancer. Does he reach in to tweak the DNA of every cell in a tumor? Does he cut off its blood supply? And if he does cure cancer, what forces of nature does he use? Electromagnetism? The weak nuclear force? Over the past decade theists have been holding conferences, publishing papers, and writing books employing the latest findings from science in an effort to answer the scientists' curiosity, and Stenger addresses these one by one, showing that none of these sciences provides an opening for God to act in our world. Indeed, quite the opposite, as all they show—if they show anything at all about our mundane lives—is that so-called miracles and other alleged divine actions are better explained by probabilities and the operation of chance than by Someone Up There running the show.

What I love most about the writings of Victor Stenger—so well exemplified in *Quantum Gods*—is that he doesn't mince words or pull punches. He isn't disrespectful and he never dissembles, but neither does he waste anyone's time by skirting around the central tenets of claims and arguments made for the existence of Something Else that science has yet to discover. Moving past all the traditional theists, something in the range of 10 to 20 percent of the population believes in some other supernatural or paranormal or spiritual force or entity or being "out there" somewhere, and in *Quantum Gods* Stenger is not about to let them off the hook just because they don't believe in Yahweh or call themselves Christians. Either there is evidence for the supernatural and the paranormal, or there isn't. There isn't. Victor Stenger explains why there isn't. Read this book to find out why.

PREFACE

The content of my previous book *God: The Failed Hypothesis* was encapsulated in its subtitle: *How Science Shows That God Does Not Exist.* In that book I tried to be very clear that I was not talking about every conceivable god, just the God with a capital "G." This is a God who not only created the universe but continues to play a central role in its operation and, most important, in the lives of humans for whom he has reserved a special place in the scheme of things.

I used the traditional Judeo-Christian-Islamic God as understood by the great mass of his worshippers rather than as understood by a handful of theologians and apologists as my model. This is a personal God who intervenes regularly in the workings of the world and in the lives of humans, performing miracles such as seeing that a certain favorite survives a plane crash or a wartime bomb misses a cathedral. This God is responsible for every leaf falling to the ground and listens to every human thought.

I argued that the actions of such a God in the physical world should surely be detectable by both the human senses and the scientific instruments that extend the range of those senses. Using the scientific method of hypothesis testing, I provided evidence beyond a reasonable doubt falsifying such a God. In doing so I independently confirmed the conclusion of many philosophers that a God with certain attributes, in particular omnibenevolence, omnipotence, and omniscience, is logically impossible given the world as we know it.

I am not finished. Other conceivable gods can be imagined whose attributes also lend their actions to be examined under the light of reason and science. Furthermore, even among those who hold no belief in a personalized supreme being we find the widespread conviction that "there has to be something out there" beyond the material universe. Surely everything, they say, especially life and humanity, cannot simply be particles moving around in empty space. Many people even in nations less religious than America believe there must be more to the universe than matter.

I will focus on two concepts of realities thought to lie out there beyond the material world of science that at the same time are based ostensibly on scientific principles—specifically quantum mechanics. Quantum mechanics is the early twentieth-century theory of matter and light whose development enabled physics to move from the familiar, commonplace "macro" world described by Newtonian mechanics into the mysterious, exotic atomic and subatomic "micro" world that lies beneath.

The first concept, which I term *quantum spirituality*, asserts that quantum mechanics has provided us with a connection between the human mind and the cosmos. The second concept, which I term *quantum theology*, argues that quantum mechanics and chaos theory provide a place for God to act in the world without violating his own natural laws.

Dealing with phenomena beyond normal experience, we should not be surprised to find that much in the quantum world defies common sense. Even Einstein was troubled by what he called the "spooky" aspects of quantum theory, although he made a vital early contribution to the subject when he proposed in 1905 that light is composed of material particles we now call *photons*. The particle nature of light appeared to contradict the well-known fact that light also exhibits wavelike properties and is well described as an electromagnetic wave. We find here the first example of the schizophrenia associated with quantum phenomena called the *wave-particle duality*.

Whether or not an object is a particle or a wave seems at first glance to depend on what you decide to use as the measure. Those eagerly looking for something else out there, something with human qualities, have seized upon wave-particle duality as implying that reality itself is a product of human consciousness. The primary theme of the quantum spirituality movement is that "we make our own reality." This principle is the subject of many books in which the authors grandly claim a new "paradigm" in our understanding of the nature of reality, with the human mind somehow tuned into a "cosmic consciousness" that pervades the universe. Based on this notion, self-help gurus offer healing therapies alternative to those of mainstream medicine, none of which are verified by clinical studies but still rake in the dollars of people desperate for answers that conventional science cannot provide.

Quantum spirituality has been linked to Eastern mysticism, so it finds itself a welcome audience, at least in America, of Buddhists and Hindus along with the various counterculture groups who have been predicting the dawning of a "new age." Although at least two decades old, the new age should now be reaching fruition, as with the turn of the millennium we have finally entered the astrological Age of Aquarius.

The second new imagined reality I will consider is a God who created the universe but does not act in any way that is inconsistent with the laws of nature. This God would be very difficult to detect by the means I applied in *God: The Failed Hypothesis.*

Such a God would not perform miracles, where a miracle is defined as a violation of a law of nature, so we would not expect to see any miracles. We don't. Such a God would have left no evidence behind at the creation, so we would expect creation to appear perfectly natural to physicists and cosmologists. It does. Such a God would make sure all his designs in nature showed no signs of that design, so we would expect living organisms and their planet to appear undesigned to biologists and geologists. They do. Such a God would not answer prayers, so we would not expect to find any evidence that prayers are answered. We don't. Such a God would not reveal facts to humans that they cannot have obtained by sensory means, so we would not expect to find any evidence for such revelations. We don't.

In *God: The Failed Hypothesis* I considered the case in which God deliberately hides himself. Such a God rewards those who have faith despite the absence of evidence, while damning those who honestly wish to believe but simply cannot do so without any evidence. I asserted that this is not a moral, beneficent God. A moral God who deliberately hides himself, exacting punishment on those who do not believe for good reasons while favoring just those few who believe for no good reason, is logically impossible. The very existence of nonbelievers in the world who are open to evidence for God proves that such a God does not exist.

However, we can imagine a God who deliberately hides from us but issues no punishment (or reward for that matter) if we fail to believe in him and no reward (or punishment for that matter) if we do. Such a God would not need his creations to grovel before him. Indeed, why would he if he is perfect and already enjoys infinite gratification?

Theologians have grappled for centuries with the logical inconsistency between a perfect, omniscient God who knows everything that is going to happen—indeed makes everything happen—and one who still allows human free will. Without free will there is no sin and atonement, contradicting one of the most fundamental tenets of Judaism, Christianity, and Islam.

For a brief period encompassing the eighteenth century and not much more, a handful of thinkers in Europe and America broke openly with Christian teaching—Protestant and Catholic—and proposed the existence of a god who created the universe but left it alone thereafter. Isaac Newton had just introduced his laws of motion, which implied that the universe was a vast machine, a clockwork in which everything that happens is predetermined by what goes on before. This new theory of god was called *deism*. For the reasons given above, such a god is probably undetectable, so long as he saw that the creation itself broke no natural laws in any detectable way.

Deism made sense to these scholars who were bent on applying reason to every aspect of life. If god were perfect, why would he need to step in after he had created the universe and its laws? The primary founders of the American republic were deists and the case can be made that the republic was not founded on Christian principles, as is so often asserted, but on deist principles.

However, deism barely lasted the century as a recognizable belief system. In the meantime, the religion of the general populace in America and Europe paid no attention to reason and focused on feelings and emotion as the means of spiritual fulfillment.

This, however, did not eliminate the theological problem of reconciling the traditional hands-on God, who allows human free will within the Newtonian clockwork universe, where his own laws already determined at the creation everything that happens, including human behavior. The deist argument, though not often acknowledged as such, may be more resilient than anyone realized because it makes so much more sense than the God religious leaders impose on their members. As we will see in chapter 1, one of the big surprises of a recent survey examining American religious beliefs is that 44 percent of Americans do not believe in a God who acts in the world or in their personal lives. That is, they may be better classified as deists rather than Christians.

Modern Christian theologians fully recognize the problem of finding a place for God to act in the world while still being consistent with science, that is, where God does not indulge in miracles that violate the laws of nature. Their problem involves not only physics but also biology and neuroscience.

The Darwin-Wallace theory of evolution is based on a combination of random mutations and natural selection. This implies that humanity is the result of countless random events from the origin of life on Earth itself over three billion years ago to the appearance of *Homo sapiens* at an estimated two hundred thousand years ago. If, as most religions preach, God created the universe with a special place for humanity in mind, and if some theologians are to remain consistent with the overwhelming scientific consensus in favor of evolution, then they have to find a way for God to have acted during the course of evolution to guarantee humans specifically evolved, not simply appeared as some random intelligent life-form.

Finding a place to act in the world of physics is a somewhat easier problem, but still not a sure thing. First, in the early twentieth century, quantum mechanics indicated that physical phenomena are not fully determined by Newton's laws of motion and the physical universe is basically indeterministic at the atomic and the subatomic levels. Second, *chaos theory* has indicated that much (but not all) that happens on the human scale is in practice unpredictable. We will discuss the attempts to solve the problem of God's action by the use of quantum mechanics and chaos.

We will find there is no escaping a large element of randomness in the universe over which either any existing god has no control or actually utilizes as part of its plan for the universe. It will turn out that only one possible god exists, the god that Einstein deeply opposed—the god who plays dice. We will see that modern cosmology indicates that at its earliest definable moment the universe was very possibly in a state of complete disorder and so retains no memory of anything that went before, including intentions of any creator. The universe looks just as it should look if it is composed of matter that appeared out of nothing. It may have been created by a god who plays dice, but that god produced a universe in which he plays no role and might as well not exist.

While I will present the picture of the material universe with its origin and laws, I will not argue passionately with those who insist that it is more than simply "particles in motion." I agree with a number of authors who have written about a concept called *emergence* in which new principles of material behavior arise as the large number of particles that make up material systems become increasingly complex. However, I do not go so far as some in claiming that some emergent principles are endowed with a property called *top-down causality*, whereby actions occur in the opposite direction to the *bottom-up causality* that physicists use to describe the universe in terms of elementary particles. That is, systems of higher complexity are said to be able to make fundamental alterations to lower-level systems.

As I will show, no convincing observations demonstrate top-down causality. I will provide examples from the first level of emergence, when thermodynamics and fluid mechanics emerge from particle physics, which indicates the whole process is still perfectly natural and material and that the only existing top-down causality is as trivial as causing the particles in a wheel to move in a circle by spinning the wheel. In short, emergence is real. But the whole is still the sum of its parts.

The chapters in this book alternate between discussions of the various claims of quantum spiritualists and quantum theologians and a survey of what twentieth-century physicists really said about their scientific claims. The physics chapters may be difficult going for those not familiar with modern physics, although they are written in layperson's terms without equations. I have deliberately tried not to oversimplify these discussions because much of the confusion that exists, not so much with highly trained theologians but with the less sophisticated gurus of the new spirituality, is the result of misunderstanding and, in some cases, intentional misrepresentation. I hope the reader will persevere since science and the ability to think critically, inside or outside science, can be learned only by diligent effort.

Although most of this book contains new material, I have had to repeat several arguments that were presented in some of my earlier books. I hope the reader will not consider me immodest in providing references to my own published work. However, I cannot expect most readers to go back through those references, so I am forced to discuss several arguments again for the sake of completeness. Rest assured that I have not simply copied what I have previously written. I have not even gone back to read my original offerings but have started each argument over again from scratch. I expect that each time I do this it gets clearer since I do have a good memory for the questions and comments of earlier readers and have taken these into account in the new text.

I have provided as complete a set of references as possible and attempted to quote other authors precisely and in context. I do not like it when people refer to my own work without quoting me but instead lazily put their own words in my mouth, almost always inaccurately representing what I really said. So I have taken pains not to do that with those whose ideas I am challenging here, where I either provide direct quotations or precise paraphrases that do not intrude my own interpretation of the words.

I have also provided World Wide Web addresses where I have found some of my references. The reader is cautioned that many of these links are temporary and may no longer be active when she tries them. In that case, a simple Web search will often turn up a new link for the same material, as well as other related matter.

You are welcome to e-mail your comments to me. My current address is vic.stenger@comcast.net. You may also find my extensive Web site helpful: http://www.colorado.edu/philosophy/vstenger. Again, if you run into broken links, do a Web search on my name.

ACKNOWLEDGMENTS

As in my other books, I have relied heavily on feedback from the e-mail discussion list avoid-L@hawaii.edu. I am especially grateful to Bob Zannelli for taking over the burden of managing this list for me as well as for his many helpful comments and suggestions on this manuscript. Thanks also to John Anderson, Eleanor Binnings, Lawrence Crowell, Keith Douglas, Yonatan Fishman, Geoff Gilpin, Alan Grayson, Bill Jeffreys, Bruce Kellett, John Mazetier, Don McGee, Brent Meeker, Jan Willem Nienhuys, Anne O'Reilly, Christopher Savage, David Sheffield, Tore Skogseth, Jeff Thompson, Imants Vilks, and Ed Weinmann. And, as always, I could not function without the loving support of my wife, Phyllis, daughter, Noelle Green, son, Andy, son-in-law, Joe Green, and daughter-in-law, Helenna Nakama. And bringing the joy to my life that makes it all worthwhile are my grandchildren, Katie Stenger, Lucy Green, Zoe Stenger, and Joey Green.

1

BELIEF AND NONBELIEF IN AMERICA

The Government of the United States of America is not in any sense founded on the Christian religion.

—Treaty of Tripoli, ratified by the US Senate and signed by President John Adams in 1797

RELIGION AND WEALTH

America is certainly an anomaly when it comes to religion. A recent report by the Pew Research Center studied the relationship between a nation's religiosity and its wealth as measured by standardized per capita gross domestic product (see figure 1.1).[1] Pew defined religiosity using a three-item index ranging from 0 to 3, with 3 representing the most religious position. Respondents were given a 1 if they believed faith in God is necessary for morality; another 1 if they said religion is very important in their lives; and a 1 if they prayed at least once a day.[2]

A clear negative correlation between religiosity and wealth is seen for most countries, the curve on the figure representing an average over all countries. At the low end of the wealth scale and high on the religiosity scale are the countries of Africa. Near the opposite end of the wealth scale,

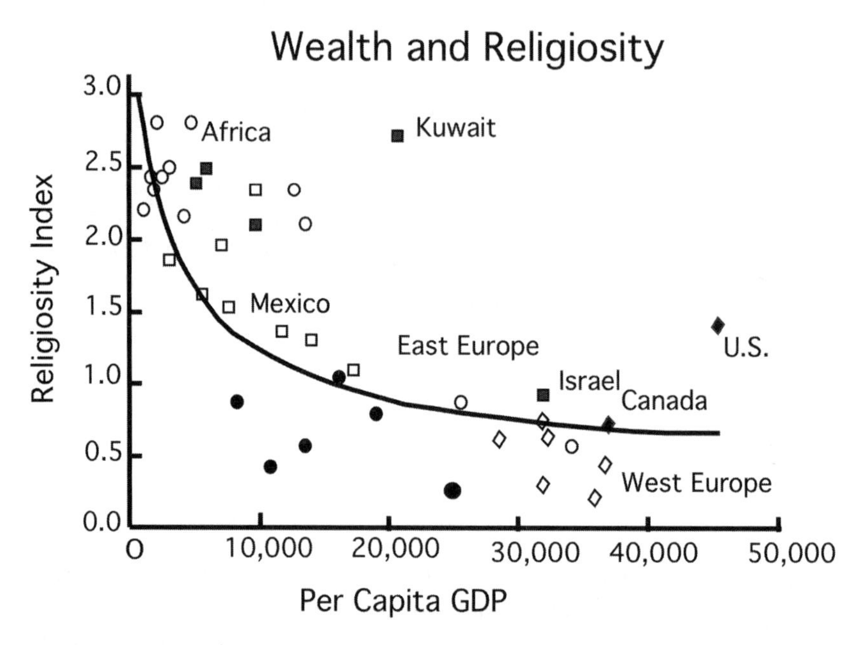

Fig. 1.1. Relationship between a nation's religiosity and its wealth. Reprinted by permission of the Pew Global Attitudes Project.

falling just a bit below the curve, are the nations of western Europe. And way out at the end of the wealth scale but well above the religiosity curve we find the United States.

Americans seem unusually religious. However, we need to examine the types of religious beliefs in America and look at the latest trends.

NO LONGER PROTESTANT

Another Pew study published in 2008 found that 28 percent of American adults had left the faith in which they were raised in favor of another religion or no religion at all.[3] The number is even greater, 44 percent, when switching from one form of Protestantism to another is included. This alone indicates some measure of turmoil among religious Americans.

Perhaps the most significant result of this survey, which is backed up by other surveys, is that America is rapidly losing its Protestant majority. In 1993 the nation was 61 percent Protestant. By 2006 this had slipped to 50 percent. Catholics have held steady at about 25 percent, but this is attributed to the large flux of immigrants from Latin America who now constitute almost half of all Catholics in the United States. In fact, more than 10 percent of Americans raised as Catholics have left the faith.

The Pew survey lists 16.1 percent of Americans as unaffiliated with any religion, breaking down into 1.6 percent calling themselves atheists, 2.4 percent agnostics, and 12.1 percent "nothing in particular." The latter group is further broken down into 6.3 percent "secular unaffiliated" and 5.8 percent "religious unaffiliated."

THE NATURE OF AMERICAN BELIEFS

The Pew results are largely consistent with an extensive study on the nature of American religious beliefs conducted in 2005 by Baylor University. The overall results based on the respondents own statements of religious preference are summarized as follows:

Evangelical Protestant	33.6 percent
Mainline Protestant	22.1
Catholic	21.2
Black Protestant	5.0
Jewish	2.5
Unaffiliated	10.8
Other	4.8

Table 1.1. Religious preferences of Americans (2005)[4]

Note that Pew gives 16.1 percent as unaffiliated. The difference with the above survey may be in the 5.8 percent who, in the Pew survey, called themselves "religious unaffiliated." In any case, those unaffiliated with any religion are not all atheists or agnostics. In the Baylor survey, 62.9 percent of the unaffiliated say they believe in God or "some higher power." While about a third say that they pray, nine out of ten never attend church services.

While most respondents agreed that God exists, they differ widely on their ideas of God, the paranormal, and religious practices.

The investigators found that they could divide their subjects into believers who followed four different types of gods, with the remainder being nonbelievers:

Type A: Authoritarian God (31.4 percent): Individuals who believe in the Authoritarian God tend to think that God is highly involved in their daily lives and world affairs. They believe that God helps them in their decision making and is also responsible for global events such as economic upturns or tsunamis. They also feel that God is quite angry and is capable of meting out punishment to those who are unfaithful or ungodly.

This is the largest belief group in all demographic categories except those with household incomes greater than $100,000 and those with a college education. The numbers fall off slightly with age.

Type B: Benevolent God (23 percent): Like believers in the Authoritarian God, believers in the Benevolent God tend to think that God is very active in their daily lives. But these individuals are less likely to believe that God is angry and acts in wrathful ways. Instead, the Benevolent God is mainly a force of positive influence in the world and is less willing to condemn or punish individuals. This belief group is twice as high among females as males, higher for whites than African Americans.

Type C: Critical God (16 percent): Believers in a Critical God feel that God really does not interact with the world. Nevertheless, God still observes the world and views its current state unfavorably. These individuals feel that God's displeasure will be felt in another life and that divine justice may not be of this world. This belief group is high among African Americans but low among whites, those with a college education, and those with higher incomes.

Type D: Distant God (24.4 percent): Believers in a Distant God think that God is not active in the world and not especially angry either. These individuals tend to think about God as a cosmic force who sets the laws of nature in motion. As such, God does not "do" things in the world and does not hold clear opinions about our activities or world events. This belief group is highest for those with a college education and those with household incomes of more than $100,000 per year. It is the belief of 28 percent males, about the same as Type A, but only 3.4 percent of African Americans.

Atheists (5.2 percent): Atheists are certain that God or gods do not exist. Nevertheless, atheists may still hold very strong perspectives concerning the morality of human behavior and ideals of social order, but they have no place for the supernatural in their larger worldview. The survey found negligible atheist African Americans, 7.8 percent males compared to only 2.7 percent females, 6.7 percent among college educated, and 6.2 percent of those with high incomes.

The above definitions are all from the survey. In the Type D case, I don't think they meant to imply that God does not hold clear opinions but that it is not clear that he cares about our activities.

Remarkably, these results indicate that many people who think of themselves as Christians disagree with basic Christian teachings. Of the four types of belief defined above, only the Type A Authoritarian God seems to be strictly traditional Christian, with Type B Benevolent God probably still consistent with general Christian teachings. The rest do not hold traditional Christian beliefs.

THE DEIST GOD

If we combine the Type C Critical God and the Type D Distant God with atheists we may have almost half of all Americans, 45.6 percent or 137 million people, who either do not believe in God or believe in a God who does not act in the universe except setting it going on its way according to natural laws he created. He seldom or never interferes with the world, in this view. This result agrees with a 2006 Harris poll, which found that 44 percent of American adults believe that God "observes but does not control what happens on Earth."[5]

The Type C and D gods are far closer to the deist god of the eighteenth-century Age of the Enlightenment than the Christian God. Now, I am sure most believers in these gods still regard themselves as Christians (as did many Enlightenment deists). Nevertheless, we can safely label as deist anyone who believes in a god who created the universe but plays no further role in it.

Even before this book appeared in print I was criticized for attempting to define Christianity, admittedly a difficult task even for a theologian. Please note that I am doing no such thing. I am not saying what a Christian is. I am simply saying what a Christian is not. As we will see in later chapters when we get deeper into theology, someone who does not believe that God acts daily to heal and empower human lives is not a Christian.

Although no denominations today except perhaps Unitarian Universalism associate themselves with deist thinking, this was once a very respectable view and has probably existed all along, unrecognized, since the Enlightenment. In chapter 4 we will discuss the rise of deism and see how it became a popular belief among intellectuals in the eighteenth century, including many of the Founding Fathers of America. Indeed, our first four presidents all seemed to share some form of deist belief. Jefferson's "creator" in the Declaration of Independence was the deist god.

As we move to the modern world we will find that the original concept of a deist god is no longer compatible with existing scientific knowledge and we will investigate how deism must be updated to be consistent with that knowledge. Several prominent Christian theologians and scientists today will be seen to be moving toward a modern form of deism that requires them to do some mighty logic twisting to still call it Christian.

PARANORMAL BELIEFS

The Baylor study also examined the relationship between religious belief and belief in what is generally termed the *paranormal*. These are supposed phenomena that lie outside conventional science. Subjects were asked about their attitudes toward ancient civilizations (Atlantis), alternative medicine, astrology, psychic phenomena, haunted houses, Ouija boards, prophetic dreams, UFOs, and strange creatures (Bigfoot, Loch Ness Monster).

The survey found that almost 76 percent of Americans believe that some alternative treatments are at least as effective as conventional medicine. Over 50 percent believe dreams can sometimes foretell the future. Over 37 percent believe houses can be haunted, and 28 percent think that the world can be influenced directly by the mind.

Those who investigate paranormal phenomena disagree on whether they necessarily involve supernatural forces. But at least some such phenomena, if confirmed, would be difficult to explain by natural forces alone.

Many scientists argue that science has nothing to say about the supernatural and so should not even get involved in such cases. The claim is that science, by definition, only deals with the "natural." I emphatically disagree. Whether any of these proposed phenomena are of supernatural origin or not, they all involve physical observations and so can be empirically studied by normal scientific means. Let me give an example.

A common claim among self-proclaimed *psychics* is that they can tell the future. This can easily be tested by some prediction coming true. Now, these can't be obvious predictions like an earthquake will strike Los Angles someday. Predicting the exact day well before the actual event would be positive evidence for psychic powers, although one or two additional successful predictions of this type by the same psychic would be needed to be sure it wasn't a lucky guess. Needless to say, no such successful prediction has ever been recorded.

For the most part, all of the phenomena listed in the Baylor study have been tested in controlled scientific experiments and have failed to be confirmed at a level where they have become accepted scientific knowledge.[6] If they had, they would be normal rather than paranormal. This includes all of alternative medicine, which fails to be confirmed as effective other than as a placebo despite widespread testimonials and belief even among health professionals.

The Baylor survey found another interesting result. An inverse relationship exists between church attendance and paranormal belief, with those who attend church weekly rating the lowest in paranormal belief and those who attend rarely or never rating the highest. However, note that the list used to define paranormal beliefs does not include any that are normally associated with religion, such as belief in demons or mystical revelations. So I would not read into this that religion provides any kind of shield against paranormal belief. The safest conclusion is that most people believe there is more to existence than what we see with our own two eyes and what science reveals with its most powerful instruments. Churchgoers just share a different set of paranormal beliefs than nonchurchgoers.

SOMETHING OUT THERE

Perhaps three out of four Americans believe that there is "something out there" and if it is not the God of religion, deist or theist, then it is still some cosmic force that acts outside the range of both normal human experience and conventional science, that is, a divine intelligence—the ground of all being.

Somewhat over 50 percent of the population believes in an authoritarian, creator God who plays a dominant role in the universe—the God of the three great monotheisms. They are *theists*. Of these, 30 percent also believe that this God plays a dominant role in human lives. In my previous book *God: The Failed Hypothesis*, I argued that such a God should have been detected by now, by science if not common experience, and so can be shown not to exist beyond a reasonable doubt.[7] This also agrees with the conclusions of many philosophers that such a God is a logical impossibility.[8]

These arguments still hold but with perhaps slightly less force for the 20 percent who think God does not participate significantly in their own lives. They still believe in a divine creator, an intelligent designer. In my book I showed that the universe looks very much as it should look if it were not created or designed but appeared and evolved by natural forces alone. I also showed that the universe looks very much as it should look if there were no creator who designed the universe with a special place for humanity. In these cases I demonstrated that a God with these properties also would be empirically detectable and that the lack of evidence that should be there is sufficient to falsify the hypothesis of that God.

Now, the 44 percent of the population whom I have labeled (without their knowledge or approval) *deists* believe in a creator that is more difficult to rule out scientifically. Their creator authored the laws of nature but then let the universe carry on according to those laws, never stepping in to change anything. For now let us assume this is a possible god, but we will see later that the only viable deist god may be one that few deists from the Enlightenment to the present really have in mind.

This leaves atheists and about the 5–7 percent or so of Americans who are unaffiliated with any church and yet still believe in a higher power. The latter also think "there must be something out there," but it is not the god of either theists or deists.

I surmise that this group is very sympathetic to a variety of ideas now in the marketplace that are labeled "spiritual." Evidence for this can be found in the sales strength of books and popularity of films that tout the notion of a *cosmic consciousness* that pervades the universe that includes the human mind. Since much of that movement is linked to Eastern philosophy and mysticism, we can add to the sample a good fraction of those who call themselves Buddhists or Hindus.

This "New Spirituality" (or "New Age Spirituality") will be the second set of beliefs, along with modern deism, that will be discussed in this book. We will see that physics, in particular quantum mechanics, plays a big role in the theory behind both these movements.

One of the most important developments in religion in America in recent years has been the "megachurch," a huge facility with a weekly attendance of thousands. The megachurch provides almost everything an individual or a family needs for their social and religious lives. The large sanctuary has the finest in visual and sound effects and the music is lively and popular. The sermons typically are the very opposite of hellfire and brimstone but promise an easy life and prosperity just by following Jesus. And almost everybody (in their community) goes to heaven.

The *New York Times* described Southeast Christian in Glendale, Arizona, as "a full service '24/7' sprawling village, which offers many of the conveniences and trappings of secular life wrapped around a spiritual core. It is possible to eat, shop, go to school, bank, work out, and scale a rock-climbing wall—all without leaving the grounds."[9]

While most megachurches are evangelical and almost all Christian, Lakewood Mile Hi Church near Denver, Colorado, seeks to blend science and religion. With 5,500 members, a modern auditorium, loud contemporary music, jumbo screens, a media store, and a childcare center, Mile Hi resembles a typical Christian megachurch. However, instead of Christianity or any other familiar religion, Mile Hi teaches the "science of mind and spirit." According to senior minister Roger Teel, "The ultimate truth for us is that we live in infinite love and oneness. We are expressions of the divine."[10]

SOMETHINGISM

As we saw in figure 1.1, both eastern and western Europe lie below the line that represents the average religiosity of the world's nations expected for their given wealth. The contrast with the United States is particularly remarkable given the otherwise close similarity in cultures. However, as with all surveys you have to look at the questions asked.

A survey of European Union nations published in 2005 concluded that only 52 percent of EU residents believe in God.[11] However, an additional 27 percent said they believed in "some spirit or life force," while 18 percent did not believe in either. In the United Kingdom these three numbers are: 38, 40, and 20; in France: 34, 27, and 33; in the Netherlands 34, 37, and 27. Clearly Europe has more atheists than the United States. But the god-belief comparison is not so clear.

The American surveys that I looked at did not always distinguish between God and the "something-else" supernatural. If we look again at the Pew survey and assume that the 16.1 percent of Americans who are unaffiliated with any church do not believe in any traditional God, then that puts about 84 percent of Americans believing in some kind of God. If we then subtract the 44 percent of Americans who seem to be deists rather than theists—even though they may still go through the motions of theism—this leaves about 40 percent of Americans actually believing in the God of their fathers. I admit this is speculative, so I will not insist on any exact numbers here and simply suggest that Americans and Europeans may not be so far apart in their God beliefs as usually assumed.

They also may not be too far apart in the non-God category of belief in "something out there." In fact, they have a name for it in Europe. In the Netherlands they call it *Ietsisme*, which translates from Dutch to English as *somethingism.*

While I estimated above that only about 5 to 7 percent of Americans are "somethingists," this is probably an underestimate. Of course, we will not really be able to make a good comparison between Europe and America until we have surveys of both asking the same questions. In particular, we need to know the fraction of European God-believers who are deists and the number of American nonaffiliates who are somethingists. Also, it seems reasonable that many people I have labeled "deists" do not hold strong beliefs in a god specifically but would prefer, if given the chance, to say that they are not sure about god and that their strongest conviction is "something must be out there." In this book I look at the scientific evidence that should exist in that case.

NOTES

1. Pew Global Attitudes Project, "World Publics Welcome Global Trade—but Not Immigration," Pew Research Center, October 4, 2007, http://pewglobal.org/reports/pdf/258.pdf (accessed May 12, 2008).

2. Ibid., p. 4.

3. US Religious Landscape Survey, Pew Forum in Religion and Public Life, Pew Research Center, 2008, http://religions.pewforum.org/reports (accessed May 12, 2008).

4. Baylor Religion Survey, Baylor Institute for Studies of Religion, September 2006. Selected findings at http://www.baylor.edu/content/services/document.php/33304.pdf (accessed May 12, 2008).

5. Harris Poll #80, October 21, 2006, http://www.harrisinteractive.com/harris_poll/index.asp?PID=707 (accessed March 23, 2008).

6. Victor J. Stenger, *Physics and Psychics: The Search for a World beyond the Senses* (Amherst, NY: Prometheus Books, 1990).

7. Victor J. Stenger, *God: The Failed Hypothesis—How Science Shows That God Does Not Exist* (Amherst, NY: Prometheus Books, 2007).

8. Michael Martin and Ricki Monnier, eds., *The Impossibility of God* (Amherst, NY: Prometheus Books, 2003).

9. Patricia Leigh Brown, "Megachurches as Minitowns," *New York Times*, May 9, 2002.

10. Electra Draper, "Megachurch Has the Frills but Doesn't Fit the Mold," *Denver Post*, May 26, 2008.

11. "Social Values, Science & Technology," *Special EUROBAROMETER* 225, June 2005, http://ec.europa.eu/public_opinion/archives/ebs/ebs_225_report_en.pdf (accessed March 22, 2008).

2

WHAT THE BLEEP IS THE SECRET?

The physical world is a creation of the observer.

—Deepak Chopra[1]

AMANDA IN QUANTUMLAND

In 2004 an independent documentary film appeared in theaters around the country called *What the Bleep Do We Know!?*[2] According to the film's Web site, it "went on to become one of the most successful documentaries of all time... while serving up a mind-jarring blend of Quantum Physics, spirituality, neurology and evolutionary thought." Modestly funded, it grossed $10 million with more in spin-offs.

In the film, Amanda, a deaf photographer played by Marlee Matlin, finds an Alice in Wonderland world of quantum uncertainty hidden behind familiar reality—all dramatized by extensive special effects. The theme is simply stated: *Quantum mechanics teaches us that we make our own reality.* As we will see, this theme is central to what I call *quantum spirituality*, going back to the 1970s to an era that was called the *New Age.*

Interspersed with Amanda's experiences in the film, many of the individuals who have been promoting quantum spirituality over that period are interviewed. These include a number of PhD physicists who have written books suggesting a connection between quantum physics and consciousness—Fred Alan Wolf, the author of many popular books on quantum spirituality; John Hagelin, prominent leader in the *transcendental meditation* (TM) movement; and Amit Goswami, retired professor from the University of Oregon and author of *The Self-Aware Universe.*[3] Appearing in an expanded later version, or "director's cut," is David Albert, who directs a program in the philosophical foundations of physics at Columbia University.

Also interviewed are two physicians who have written extensively about quantum mechanics and consciousness, anesthesiologist Stuart Hameroff and psychiatrist Jeffrey Satinover, along with assorted spiritualists, and finally, a 35,000-year-old warrior.

In this chapter I will briefly introduce quantum spirituality by providing quotes from the *Bleep* interviewees and their published writings. I will also relate the similar claims made in an even more successful documentary released in 2006 called *The Secret*,[4] along with some of the statements by earlier proponents of quantum spirituality. I will simply report and for the most part not evaluate these views at this time. In the chapters to follow we will discuss in nontechnical detail what quantum mechanics teaches that has led many of those seeking new paths to enlightenment, supported by a small minority of highly trained scientists, to reach revolutionary conclusions that go well beyond the mainstream of physics. I will then ask if these conclusions are justified by either the theories of physics or the empirical facts.

The first BLEEPER I will quote is Fred Alan Wolf. Wolf received a PhD in physics from the University of California at Los Angeles (UCLA) in 1963. So did I! I don't recall ever meeting him. His *Taking the Quantum Leap* won the 1982 National Book Award for Science.[5] Wolf, who fashions himself as "Dr. Quantum," has appeared on many radio and TV shows, including the Discovery Channel, and should be regarded as a popularizer rather than a serious scholar. Here's how Wolf summarizes the role of physics in consciousness:

> The importance of consciousness as an element in physics is becoming apparent.... Consciousness acts or has an effect on physical matter by making choices that then become manifest. It now appears that such an

> action cannot simply take place mechanically. Implied now is a "chooser," or subject who affects the brain and nervous system.... I suggest that this chooser/observer does not exist in spacetime and is not material, which suggests that it is a spiritual essence or being residing outside of spacetime.[6]

John Hagelin ran for president on the Natural Law Party ticket in 1992, 1996, and 2000. We will meet him again later when we discuss Maharishi Mahesh Yogi and the transcendental meditation (TM) movement.

According to Hagelin,

> The quantum-mechanical Unified Field of Natural Law is a field of self-referral consciousness which generates the whole manifest universe by its process of self-observation.[7]

David Albert does not appear in *Bleep* but is the first one interviewed in the expanded director's cut sequel, *Down the Rabbit Hole*.[8] He was "outraged" when he saw the product and disassociated himself from the project. He says he told his interviewers that quantum mechanics has nothing to do with consciousness and spirituality, but the filmmakers rearranged his words to make it sound he as if he were saying the opposite.[9]

In *Beyond the Bleep: The Definitive Unauthorized Guide to* What the Bleep!? Alexandra Bruce identifies Amit Goswami as the one physicist interviewed who "expresses views which are so antithetical to what is accepted by both Western science and common sense that his statements have become a lightning rod for the film."[10] I agree that Goswami's thinking accurately represents the supposed scientific and philosophical basis, such as it is, for quantum spirituality, so let us look at his views in a little more detail.

THE SELF-AWARE UNIVERSE

Amit Goswami received a PhD in physics from Calcutta University in 1964 and taught physics at the University of Oregon for thirty-two years. In the 1980s Goswami proposed an interpretation of quantum mechanics he called *monistic idealism*.

Monism is a name usually attached to the philosophical doctrine in which the universe is composed of one kind of stuff. For example, material monism assumes the world is made of matter and nothing else. This is to be compared with *dualism* in which, for example, in Cartesian dualism, the universe is composed of matter and some other component such as "spirit" or "soul." I think it would be fair to say that most of the world's religions teach some kind of Cartesian dualism, including Hinduism, the main religion in Goswami's nation of birth, India. On the other hand, the vast majority of philosophers of mind and neuroscientists hold to material monism and view the mind as a product of matter. This marks an even greater gulf between religion and science than evolution that has yet to have broken out into political warfare.

Goswami's monism, which he says is drawn from both the Hindu school of Advaita Vedanta as well as Theosophy, claims that the universe is not composed of matter at all but of a "universal consciousness." As Goswami puts it in his *Bleep* interview:

> The material world around us is nothing but possible movements of consciousness. I am choosing moment by moment my experience.[11]

Goswami notes that the "You" in "You Make Your Own Reality" is not the individual "you" exercising his or her free will, but the collective You (or "I") of an all-pervasive cosmic consciousness that connects all minds throughout the universe at speeds faster than the speed of light.

> As the real experiencer (the nonlocal consciousness) I operate from outside the system—transcending my brain-mind—that is localized in space-time. . . . My separateness—my ego—only emerges as an apparent agency for the free will of this cosmic "I," obscuring the discontinuity in space-time that the collapse of the quantum brain-mind state represents."[12]

The terms *local* and its opposite, *nonlocal*, will appear frequently in this book. Two events in space and time and are said to be "local" if they can be connected by a signal moving at the speed of light or less. The term *local* follows from the fact that in this case you can find a frame of reference in which the two events occur at the same place in space. For example, you get on a train at one station and get off at another, using the same door in both cases. Those two events are at different places in Earth's reference frame but at the same place in the train's reference frame. The two events are thus local because there exists at least one reference frame where they are at the same place.

It is not possible to find such a reference frame for two events that are said to be *nonlocal.* Nonlocality is usually connected with superluminality, that is the motion of bodies or signal faster than the speed of light. This is forbidden by Einstein's special theory of relativity. One of my many tasks will be to evaluate the claim that quantum mechanics is necessarily non-local, that is, involves phenomena that are superluminally connected. Einstein called these phenomena "spooky actions at a distance."

A few years before I retired from the University of Hawaii in 2000, Goswami was invited to speak to the Philosophy Department on campus and I was asked to follow his presentation with comments of my own. In those comments I accused Goswami of *solipsism,* which is the doctrine that the self is the only reality and the world is all made up in our heads. Goswami objected vehemently that this was not at all his position. However, he has said elsewhere that our notion of being separate individuals is an illusion. I still do not see how the existence of one common "self," the cosmic consciousness in which we all participate that manufactures reality, is any different from the solipsistic self who does the same. But, I suppose, the distinction does not matter. The simple thing to remember is that Goswami teaches that the universe is all in "our" heads, where "we" are the totality of universal consciousness. And how does he know this? Because quantum mechanics says it is so.

As we will see as we proceed through this book, several of the same themes are repeated over and over by the quantum spiritualists with little more than changes in language and usually not even that. "We make our own reality" is the primary theme. Another is "We are part of an inseparable whole." And "Quantum mechanics is behind it all."

OTHER BLEEPERS

Let me briefly summarize the views of two other *Bleep* interviewees who have provided input to the quantum spiritualism story.

Jeffrey Satinover is a psychiatrist with a strong physics background and a large and varied résumé. Among his several books is one called *The Quantum Brain.*[13] His main thesis is that humans are undeniably material (that is, nonspiritual) machines, but the fact that we do not behave deterministically is evidence that quantum effects must be present. As he says in *Quantum Brain*, "A 100 percent mechanical system cannot under any circumstance generate an indeterminate outcome." The fact that human beings have the capacity to act indeterministically "would either have to derive from the quantum nature of matter, amplified, or arise from some other, utterly mysterious source."[14]

Another BLEEPER who has long promoted the idea of a quantum brain is anesthesiologist Stuart Hameroff. Hameroff is noted for his collaboration with the eminent Oxford mathematician and cosmologist Roger Penrose in trying to find a place for quantum mechanics in the brain.

In 1989 Penrose wrote a best-selling tome called *The Emperor's New Mind* that was packed with wonderful material on physics, mathematics, and computers.[15] Penrose's main thesis was that the human brain is not a computer but must operate in some way that cannot be replicated with any computer no matter how powerful. That is, the brain did not follow "algorithms" in solving every problem it dealt with. He made a rather remarkable proposal that the actual mechanism had something to do with quantum gravity.

Penrose was met with considerable skepticism, especially in the artificial intelligence community—which he was basically arguing out of business—but also among physicists who could not see what quantum gravity could possibly have to do with a large, hot structure such as a brain.

Here's how Penrose and Hameroff explain their mechanism:

> According to the principles of OR (*orchestrated reduction*, proposed by Penrose in 1994),[16] superpositioned states each have their own space-time geometries. When the degree of coherent mass-energy difference leads to a sufficient separation of space-time geometry, the system must choose and decay (reduce, collapse) to a single universe state, thus preventing "multiple universes." In this way, a transient superposition of slightly differing space-time geometries persists until an abrupt quantum classical reduction occurs and one or the other is chosen. Thus consciousness may involve self-perturbations of space-time geometry.[17]

Hameroff had the idea that the quantum mechanism for consciousness involved quantum coherent effects taking place in *microtubules*, hollow cylindrical polymers that are part of the cytoskeleton of all cells in the human body, from head to toe. For some reason, only the microtubules of nerve cells in the brain participate in conscious decisions, although there is some suggestion that in the case of the human male, microtubules in the cells of the penis dominate.

RAMTHA THE LEMURIAN

Another prominent interviewee in *Bleep* is identified as Ramtha, a Lemurian warrior who fought against the people of Atlantis thirty-five thousand years ago. Ramtha is said to have conquered two-thirds of the known world before ascending to heaven at a spot near the Indus River.

On February 7, 1977, Ramtha appeared to Judith Darlene Hampton (stage name, J. Z. Knight) and her husband in the kitchen of their trailer home in Tacoma, Washington. Since then Knight has been able to "channel" to Ramtha by leaving her body and having Ramtha speak through it. Knight formed Ramtha's School of Enlightenment through which she passes on Ramtha's wisdom for a fee. She has become one of the nation's best-known spiritual mediums, earning about $10 million yearly. Knight has moved out of the trailer to a French château–styled mansion in Yelm, Washington.

The three directors of *Bleep* are all students of the Ramtha School, but they deny the charge that *Bleep* is a recruitment film for the school. In the film, Ramtha does not appear directly but is interviewed through Knight. Ramtha speaks in a deep Indian accent with an occasional hint of modern British English.

Judging from Ramtha's words, some have seen the film as an attack on Christianity and other religions. Calling Christianity a "backward" religion, Ramtha, in keeping with the quantum theme of *Bleep*, informs us that the parables of Jesus are photon waves of probability.

Here is a sample of Ramtha's teachings that relate to the quantum:

> The soul is the recorder of unfinished business, the tallier in which in the mind of God each subject's achievements are added to this fluid mind that the ancients used to call the Akashic Record, but all it means is space. And we know it today in a much more sophisticated term called the quantum field, and its spiritual name is the mind of God.[18]

THE SECRET

Hot on the heels of the success of *Bleep* has been another blockbuster independent documentary film called *The Secret*, which makes a slight but clever variation on the primary New Age theme that we make our own reality. The main character in the film is Australian television writer and producer Rhonda Byrne. She relates how her life had collapsed around her until given a glimpse of "a Great Secret—The Secret to Life" in a hundred-year-old book given to her by her daughter Hayley.[19] She traced the history of the Secret and found it was known by some of the greatest men of history: Plato, Shakespeare, Newton, Hugo, Beethoven, Lincoln, Emerson, and (of course) Einstein. Byrne wondered why everyone did not know the Secret and so began searching, finding one living master after another. These gurus were included in the film she made to reveal the Secret to humanity.

And what is this magnificent Secret? It is the *law of attraction.* As Bob Proctor, author of the best seller *You Were Born Rich*,[20] explains,

> Everything that's coming into your life you are attracting into your life. And it's attracted to you by virtue of the images you're holding in your mind. It's what you are thinking.[21]

Proctor travels the globe teaching the Secret, helping companies and individuals to create lives of prosperity through the law of attraction.

> And how do you make this happen? You think about it!
>
> If you think about what you want in your mind, and make that your dominant thought, you *will* bring it into your life.[22]

Here's how it works. Your mind is transmitting your thoughts throughout the universe. Each thought has its own unique frequency and attracts like things of the same frequency. So if your thought is one of becoming wealthy, you will attract wealth into your life.

So why isn't everybody wealthy? Because most people aren't thinking the right thoughts. As Byrne explains,

> The only reason why people do not have what they want is because they are thinking more about what they *don't* want than what they *do* want.

> Listen to your thoughts, and listen to the words you are saying. The law is absolute. There are no mistakes.[23]

This theme is repeated over and over in the film and the book, by the various masters of the Secret (who, for some reason, do not all look like twenty-year-old movie stars). These include our old friends, the presidential candidate John Hagelin and the ubiquitous Dr. Quantum, Fred Alan Wolf. You can guess where Wolf says the Secret comes from:

> I'm not talking to you from the point of view of wishful thinking or imaginary craziness. I'm talking to you from a deeper, basic understanding. Quantum physics really begins to point to this discovery. It says you can't have a Universe without mind entering into it, and that mind is actually shaping the very thing that is being perceived.[24]

John Hagelin gives us more details:

> Quantum mechanics confirms it. Quantum cosmology confirms it. That the Universe essentially emerges from thought and all of this matter around us is just precipitated thought. Ultimately we are the source of the Universe. ... So we are the creators of our own destiny, but ultimately we are the creators of Universal destiny. We are the creators of the Universe.[25]

In his three runs for the presidency, from 1992 to 2000, John Hagelin amassed a total of 232,000 votes.[26]

QUANTUM HEALING

Certainly we expect that one of the prime uses of the Secret for most people would be for their health. You are guaranteed good health if you think about it. However, the techniques taught in *The Secret* were certainly not unknown prior to its appearance in theaters and bookstores.

The 1952 publication of *The Power of Positive Thinking* by Norman Vincent Peale stayed on the *New York Times* best seller list for 186 consecutive weeks. Peale taught a kind of self-hypnosis in which by a series of constant repetitions of positive thoughts the subject would fight off any adversity. Peale did not attribute this to quantum mechanics, however.

Perhaps the first well-known spiritual healer to specifically teach "quantum healing" was Deepak Chopra, a physician who was originally a member of the transcendental meditation movement. In the 1980s he broke off and formed his own organization that specialized in Ayurvedic (ancient Indian) medicine and self-help. He is the author of many best-selling books, including *Quantum Healing* (1989)[27] and *Ageless Body, Timeless Mind* (1993),[28] that rest on the assumption that quantum mechanics enables us to make our own reality.

In June 2007, I was at the giant Book Expo America in New York City to help sell my book *God: The Failed Hypothesis*.[29] While waiting to be interviewed on TV I noticed all these people lined up in the area reserved for author signings. I asked someone in what was by far the longest queue whom he was waiting for. It was Deepka Chopra, who had just published his latest effort, *Life after Death*.[30]

ALTERNATIVE MEDICINE

Much of complementary and alternative medicine makes allusions to physics. In particular, a common thread in these therapies is the concept coming out of traditional Chinese medicine in which life is described as the product of an "energy field" that flows through the human body. The Chinese call it *qi* or *ch'i* (pronounced roughly chee), the Japanese, *ki*. In Sanskrit it is *prana*. In Chinese, *qi* also means breath.

In the ancient Western world it also was commonly believed that breath was the force of life, leaving the body upon death. In Greek *pneuma* is literally the word for "that which is breathed or blown" but was used in Stoic thought to refer to the vital spirit or soul of a person. The world for breath in Latin, *spiritus*, is also used to refer to spirit.

Modern alternative medicine has adopted many of the techniques of traditional Eastern therapies, such as acupuncture and massage, and has added a few such as therapeutic touch and chiropractic, which claim to manipulate the body's energy field. That field is variously identified with electromagnetic fields, quantum fields, or as we will see in the case of Maharishi, the *grand unified field* of physics.

I have already written extensively on the various forms of alternative medicine in books and articles and will not repeat any examples that I have previously discussed.[31] There are always many to choose from, with new ones coming along every day. My Google search on quantum healing yielded 1,490,000 hits. Here are just two random examples.

Quantum Touch: Practitioners focus and amplify the life-force energy of *qi* or *prana* by various breathing and body awareness exercises:

> When the practitioner resonates at a high frequency, the client often entrains to, or matches, the higher frequency, thereby facilitating healing using the body's biological intelligence. Life-force energy affects matter on the quantum, subatomic level and works its way up through atoms, molecules, cells, tissue, and structure.[32]

Quantum Depth Healing:

> Health does not come from a pill, a potion, a lotion, surgery, a doctor or even an herb. It comes from the body's own innate intelligence or inborn wisdom to survive and rejuvenate. Our bodies ultimately are domains of information, intelligence and energy. The very fact that our body is an assembled mass of molecules implies that we are energetic entities. Further, quantum physics discovered in the last decades that every particle of matter is associated with interaction and resonance quanta (parcels of energy) at a ratio of about 1 nucleon to 1 billion quanta. The quanta exhibit specific patterns and are susceptible to resonance. These subtle energetic configurations (bodies) can be disturbed causing unwellness and pain. Therefore, the body is a quantum mechanical device and Quantum Healing is healing the bodymind from a quantum level. That means from a level, which is not manifest at a sensory level, Quantum Depth Healing involves a shift in the areas of energy information, so as to bring about a reconstruction in an idea that has gone wrong. So quantum healing involves healing one mode of consciousness "the mind" to bring about changes in another mode of consciousness "the body." Another important point in quantum biology is that consciousness is not consolidated or focused in any one particular place. Each thought, feeling, desire, attitude, instinct, or drive you have affects your nervous system through all its organs and tissues by a group of chemicals called neuropeptides in the brain. Each thought, feeling, desire, attitude, instinct, or drive you have affects your nervous system by means of these specific messenger units.[33]

These examples and the others I have examined pretty much say the same thing we hear from Chopra, *Bleep*, and *The Secret.* They all rely on the claim that quantum mechanics has eliminated the reductionist, deterministic, and atomic doctrines based on Newtonian mechanics. Furthermore, they assert that quantum mechanics announced a connection between quantum, body, and mind that was previously unrecognized in the West but was already deeply embedded in traditional Chinese and Ayurvedic (Hindu) healing practices. These have been adopted by the New Age movement for some thirty years now and are hardly newly discovered "secrets" to those who have followed the movement over this period.

NOTES

1. Deepak Chopra, *Ageless Body, Timeless Mind: The Quantum Alternative to Growing Old* (New York: Random House: 1993).

2. William Arntz, Betsy Chasse, and Mark Vicente, *What the Bleep Do We Know!?* Lord of the Wind Films, LLC, 2004, www.whatthebleep.com (accessed May 12, 2008).

3. Amit Goswami, *The Self-Aware Universe* (New York: Penguin, 1993).

4. Drew Heriot, *The Secret (Extended Edition)*, TS Production, LLC, 2006.

5. Fred Alan Wolf, *Taking the Quantum Leap* (New York: HarperCollins, 1981).

6. Fred Alan Wolf, *The Yoga of Time Travel: How the Mind Can Defeat Time* (Wheaton, IL: Quest Books, 2004), p. 197.

7. As quoted in Alexandra Bruce, *Beyond the Bleep: The Definitive Unauthorized Guide to* What the Bleep Do We Know!? (New York: Disinformation Company, 2005), p. 153.

8. William Arntz, Betsy Chasse, and Mark Vicente, *What the Bleep!? Extended Director's* Cut, Lord of the Wind Films, LLC, 2004.

9. Gregory Mone, "Cult Science. Dressing Up Mysticism as Quantum Physics," *Popular Science*, October 2004, http://www.popsci.com/scitech/article/2004-10/cult-science (accessed May 12, 2008).

10. Bruce, *Beyond the Bleep*, p. 101.

11. As quoted in ibid., pp. 101–102.

12. Goswami, *The Self-Aware Universe*, p. 9.

13. Jeffrey Satinover, *The Quantum Brain: The Search for Freedom and the Next Generation of Man* (New York: Wiley & Sons, 2001).

14. Ibid., p. 80.

15. Roger Penrose, *The Emperor's New Mind: Concerning Computers, Minds, and the Laws of Physics* (Oxford: Oxford University Press, 1989).

16. Roger Penrose, *Shadows of the Mind: A Search for the Missing Science of Consciousness* (Oxford: Oxford University Press, 1994).

17. Roger Penrose and Stuart R. Hameroff, "Orchestrated Objective Reduction of Quantum Coherence in Brain Microtubules: The 'ORCH OR' Model for Consciousness," in *Toward a Science of Consciousness—The First Tucson Discussions and Debates*, ed. S. R. Hameroff, A. W. Kaszniak, and A. C. Scott (Cambridge, MA: MIT Press, 1996), pp. 507–40.

18. J. Z. Knight, "The Akashic Record and the Quantum Field," Fireside Series, vol. 3, no. 3, *Parallel Lifetimes: Fluctuations in the Quantum Field* (JZK Publishing, 2003–2005), http://ramtha.com/html/community/teachings/Akasha_and_Bohm.pdf (accessed May 12, 2008).

19. Rhonda Byrne, *The Secret* (New York: Attria Books, 2006), p. ix.

20. Bob Proctor, *You Were Born Rich* (LifeSuccess Products, 1997), http://www.synergylifesuccess.com (accessed July 12, 2008).

21. Byrne, *The Secret*, p. 4.

22. Ibid., p. 9.

23. Ibid., p. 12.

24. Ibid., p. 21.

25. Ibid., p. 160.

26. Wikipedia, http://en.wikipedia.org/wiki/John_Hagelin (accessed March 4, 2009).

27. Deepak Chopra, *Quantum Healing: Exploring the Frontiers of Mind/Body Medicine* (New York: Bantam, 1989).

28. Chopra, *Ageless Body, Timeless Mind.*

29. Victor J. Stenger, *God: The Failed Hypothesis—How Science Shows That God Does Not Exist* (Amherst, NY: Prometheus Books, 2007).

30. Deepka Chopra, *Life after Death: The Burden of Proof* (New York: Harmony Books, 2006).

31. Victor J. Stenger, "Bioenergetic Fields," *Scientific Review of Alternative Medicine* 3, no. 1 (Spring/Summer 1999); Victor J. Stenger, *Has Science Found God? The Latest Results in the Search for Purpose in the Universe* (Amherst, NY: Prometheus Books, 2003), pp. 268–76.

32. Quantum-Touch, http://www.quantumtouch.com (accessed May 12, 2008).

33. Quantum Depth Healing, http://elevatedtherapy.org.uk/index-page12.html (accessed May 12, 2008).

3

PURSUING THE TAO

He who pursues learning will increase every day;
He who pursues the Tao will decrease every day.

—Lao Tzu[1]

BERKELEY DAYS

Berkeley, California, in the '60s and '70s was famous for the large number of Nobel Prize winners on the University of California physics faculty. But it was even better known as the center of protests against the Vietnam War. The surrounding streets were the home of hippies and flower children who experimented with every form of deviation from the mainstream of American culture, from drugs to free love.

At the time, a young Austrian physics PhD by the name of Fritjof Capra was doing research in the San Francisco Bay area. He frequented the coffee shops of Berkeley, where above the din of rock music and through the fog of marijuana smoke he learned about Eastern philosophy and mysticism.

Capra saw strong parallels between Hinduism, Buddhism, Taoism, and modern physics, especially quantum mechanics and a new idea called *bootstrap theory* that was just coming out of Berkeley. He put his thoughts together into a best-selling book called *The Tao of Physics* that first appeared in 1975 and still can be found on the science shelves of most bookstores. The word *tao* derives from "way" or "path" and, in Chinese philosophy, it refers to the underlying organization of and the unfolding of events in the universe.

To Capra, the most important parallel was the "unity and mutual interrelation of all things... interdependent and inseparable parts of the cosmic whole." This was present, Capra argued, in quantum physics' discovery that seemingly causally separate parts of a physical system apparently affect the outcome of experiments on that system. This was at odds with the traditional reductionist view of physics in which everything is broken down into parts, like atoms, that can be studied independently of other parts.

Capra perceived an "inner fragmentation" of humanity, in which we view the world outside as a "multitude of separate objects and events." He attributed this to the spirit/matter dualism of René Descartes that divided nature into "two separate and independent realms."[2]

The worldview of Eastern philosophy was termed "ecological" by Capra (a philosophy sadly lost in the modern nations of Asia, with their traffic-clogged streets and almost unbreathable city air). He gave this as the main reason for the popularity of Eastern philosophy—at least in 1975 when environmentalism was on the rise following the antiwar, civil rights, and other idealistic movements. The mostly young people in these movements tended "to see science, and physics in particular, as an unimaginative, narrow-minded discipline which contributed to the evils of modern technology." Capra expressed his aim in writing *The Tao of Physics* was to improve the image of science by showing that "there is an essential harmony between the spirit of Eastern wisdom and Western science... that the way of the *Tao*—of physics can be a path with a heart, a way to spiritual knowledge and self-realization."[3]

Capra claimed to see this harmony in a new, holistic quantum physics that replaced Newtonian reductionism: "The constituents of matter and the basic phenomena involving them are all interconnected, interrelated, and interdependent.... They cannot be understood as isolated entities but only as integrated parts of the whole."[4]

He quotes from *The Central Philosophy of Buddhism:*

Things derive their being and nature by mutual dependence and are nothing in themselves.[5]

Capra sees this sentiment repeated by the Berkeley theoretical physicist Henry Stapp in a technical paper on quantum mechanics:

An elementary particle is not an independently existing unanalyzable entity. It is, in essence, a set of relationships that reach outward to other things.[6]

While connectedness is the main theme of *The Tao of Physics*, being repeated many times throughout, Capra perceived other parallels. One is the unity of polar opposites, as exemplified by the Asian concept of *yin and yang*. These are supposedly manifest in the apparent dual nature of matter in quantum mechanics, where an object is sometimes a particle and sometimes a wave. Capra tells the story of how Niels Bohr, one of the founders of quantum mechanics, chose the Chinese symbol *t'ai-chi* with the inscription *Contraria sunt complementa* (Opposites are complementary) for his coat of arms.[7]

Eastern philosophy, according to Capra, has always regarded that space and time "are not features of reality, as we tend to believe, but creation of the mind."[8] This contrasts with the Western view, exemplified by Plato, that space and time are expressions of, and part of, ultimate reality. Capra claimed that Einstein's theory of general relativity supports the Eastern view, with space and time simply being elements of language used by observers to describe their environment.

Certainly Einstein contributed to the notion that space and time are human inventions rather than fundamental parts of the universe when he associated time with what we read on a clock and distance with what we read on a meter stick. But this had to do with his theory of special relativity, which came out in 1905. In general relativity, published a decade later, the universe is described in terms of non-Euclidean space-time—Einstein seems to have changed his mind and began treating space and time as if they are objectively real.

According to Capra, Buddhists perceive all objects as processes in a universal flux and deny the existence of any permanent material substance. He says, "The Buddha taught that 'all compounded things are impermanent', and that all suffering in the world results from our trying to cling to fixed forms—objects, people or ideas—instead of accepting the world as it moves and changes."[9] Capra likened this to the situation in modern physics where particles produced in high-energy accelerators live for only short periods of time. This also jibes with the quantum picture of particles, even ostensibly stable ones such as protons and electrons, existing part of the time in other forms. For example, a photon is some of the time a positron (anti-electron)-electron pair.

In another parallel, Capra talks about the Eastern view that the reality underlying phenomena is a formless, empty void. However, this void is not nothingness but is full of *ch'i*, the source of all life. He equates *ch'i* with the quantum field. The fact that in quantum physics the vacuum is not always empty suggests some connection. As Capra says:

> From its role as an empty container of the physical phenomena, the void has emerged as a dynamic quantity of utmost importance. The results of modern physics thus seem to confirm the words of the Chinese sage Chang Tsai:
>
> "When one knows that the Great Void is full of *ch'i*, one realizes that there is no such thing as nothingness."[10]

Capra also noted that, at the subatomic scale, physicists find a constant creation and destruction of particles, and viewed this as a manifestation of the cosmic dance of the Hindu god Shiva, who symbolizes the cycle of birth and death.

Writing in 1975, Capra thought he saw East and West coming to together in the new *tao* of physics and mysticism.

THE DALAI LAMA AND SCIENCE

Perhaps the religious leader in the world today with the greatest respect for science is Tenzin Gyatso, the fourteenth Dalai Lama, who is the spiritual and political leader of the Tibetan people. Exiled from Tibet by the government of China since 1959, the Dalai Lama has traveled the world speaking to leaders from almost every corner of society. Many of those leaders have been renowned scientists who have tutored the Dalai Lama in science. While a willing student, he is untrained in mathematics and limited to more philosophical discussions.

The Dalai Lama is often asked in his travels what he would do if science showed any of his beliefs are wrong. His answer, "I would change my beliefs." In his 2005 book, *The Universe in a Single Atom*, he makes this more precise:

> My confidence in venturing into science lies in my basic belief that as in science, so in Buddhism, understanding the nature of reality is pursued by means of critical investigation: if scientific analysis were conclusively to demonstrate certain claims in Buddhism to be false, then we must accept the findings and abandon those claims.[11]

Like Capra, the Dalai Lama sees much in common between Buddhism and science, and between science and spirituality. He says that Buddha himself "advises that people should test the truth of what he has said through reasoned examination and personal experiment."[12]

However, according to the Dalai Lama Buddhism goes one step further than science. It also involves "contemplative investigation" and the "introspective examination if inner experience."[13]

So while the Dalai Lama is willing to accept the authority of science in its own area of external, objective observation and analysis, he clearly believes that Buddhist meditative practices open up another channel to truth beyond sensory observation. This is similar to the channel of revelation in other religions. As I have previously argued, if such a channel existed we should have evidence for it.[14] If someone returns from a "spiritual" experience with some new truth, then it should be possible on at least one or two occasions to test the validity of that truth. We do not have a single example of a revelation producing some truth about reality that could not have been learned in any other, natural way.

One of the Dalai Lama's closest physics consultants was the late David Bohm, an expatriate American physicist living in England after bravely refusing to testify against his friends before the US House Un-American Activities Committee and having his passport revoked as a result. In the 1950s Bohm had developed an alternative model of quantum mechanics that implies a holistic universe in which the motion of a particle is determined by the instantaneous influence of all the other particles in the universe. He became one of the primary proponents of quantum spiritualism and supported the notion we have seen expressed by Capra that the universe is one undivided whole.

Unsurprisingly, the Dalai Lama has found this idea very congenial to Buddhist teaching, where he asserts that one of the most important philosophical insights is the *philosophy of emptiness*: "All things and events, whether material, mental, or even abstract concepts like time, are devoid of objective, independent existence."[15] We will see what quantum physics says about this in a future chapter.

NOTES

1. Lao Tzu, *Tao Te Ching*, trans. Ch'u Ta-Kao, ch. 41, as quoted in Fritjof Capra, *The Tao of Physics* (Boulder: Shambhala, 1975), p. 39.
2. Capra, *The Tao of Physics*, pp. 22–23.
3. Ibid., p. 25.
4. Ibid., p. 131.
5. As quoted in *The Tao of Physics*, p. 138.
6. Henry P. Stapp, "S-Matrix Interpretation of Quantum Theory," *Physical Review* D3 (March 15, 1971): 1319. I wonder if he meant "analyzable" here.
7. Capra, *The Tao of Physics*, p. 160.
8. Ibid., p. 161.
9. Ibid., p. 191.
10. Ibid., p. 223.
11. Dalai Lama, *The Universe in a Single Atom: The Convergence of Science and Spirituality* (New York: Morgan Road Books, 2005).
12. Ibid., p. 24.
13. Ibid.
14. Victor J. Stenger, *God: The Failed Hypothesis* (Amherst, NY: Prometheus Books, 2007), ch. 6.
15. Ibid., p. 46.

4

THE GURU OF GUTS

All phenomena in the world are nothing but the illusory manifestation of the mind and have no reality of their own.

—Ashvaghosha[1]

THE MCDONALD'S OF MEDITATION

About the time physicist Fritjof Capra was hanging out in the coffee shops of Berkeley expounding on the principles of quantum physics he found embedded in Eastern mysticism, an Indian yogi trained as a physicist and calling himself Maharishi Mahesh Yogi was traveling around the United States and Great Britain expounding the principles of ancient Eastern mysticism he found embedded in quantum physics. Maharishi taught a meditation technique called *transcendental meditation* (TM) that he claimed enabled the participant to tune into the "cosmic consciousness," a quantum field that pervaded the universe. His primary teaching was again one we have found to be the central idea among all quantum spiritualists. Reality is not some objective state of the universe that would be out there whether humanity existed or not. Rather, it is the creation of the human consciousness. We make our one reality. How do we know? Quantum mechanics says so.

His audiences were small until August 1967, when his very last appearance on tour in London was attended by the Beatles. At the time Eastern religions were all the rage in the counterculture of Britain, even more so than in the United States. Early in 1968 the Beatles joined Maharishi at his Academy of Meditation in Shankarcharya Nagar, India. They led a large contingent that included the Rolling Stones, Donovan, Mia Farrow, and the Beach Boys. They didn't stay long and the Beatles quickly terminated their relationship with Maharishi.

Within a few years Maharishi had founded a global movement. TM became the McDonald's of meditation, marketed on almost every street corner with posters offering a free introductory class. Perhaps millions gave it a try.

One of these millions was Geoff Gilpin, a young student living in Green Bay, Wisconsin, who happened to spot the TM poster in 1973. Geoff would spend five years in the TM movement, first as a hanger-on, then a volunteer, and finally attending Maharishi International University in Fairfield, Iowa, leaving in 1978. Recently Geoff went back and spent some time in Fairfield visiting his old haunts. He has written about these experiences in a fascinating book called *The Maharishi Effect*, which I have relied on extensively for parts of this chapter.[2]

Meditation is an ancient practice in Eastern religions, going back before history. Usually it involves mental repetition of a sacred sound, word, or phrase called a *mantra* with the purpose of helping empty the mind of thought. Practitioners experience a great calm that many imagine puts them in touch with realities beyond the physical world, participating, as Maharishi puts it, in the great cosmic consciousness.

TM differs from Zen and other meditation forms in not concentrating so much on the mantra but making as little conscious effort as possible. It also does not require the strict celibate, vegetarian lifestyle of a Buddhist or Hindu monk, making it agreeable to those in the more self-indulgent West looking for something a little different in their lives.

Furthermore, studies in Fairfield and elsewhere indicated that meditation has some marginally beneficial medical effects such as lowering blood pressure. It has never been demonstrated, however, that TM is unique in this regard among meditation techniques. Independent experiments at

Harvard and elsewhere indicate that almost any relaxation method, including Jewish, Christian, and Muslim prayers and even those with no religious content, work equally well.[3] None have demonstrated major curative capacities.

But whatever the actual physical benefit, TM was sufficiently satisfying to a great many practitioners who became convinced that there was something "spiritual" to it, something that they interpreted as a mystical or religious experience. Maharishi taught that people who practice his meditation techniques can levitate, control the weather, put an end to war, and generally create heaven on earth. This was all possible because quantum mechanics was part of our consciousness. The mechanism is the grand unified field of cosmic consciousness, the long-sought goal of physics that, according to TM believers, has now reached fruition. According to the Web site of the Yogic Flying Clubs for Students, a group dedicated to the TM technique of "yogic flying,"

> Progress in theoretical physics during the past quarter century has led to a progressively more unified understanding of the laws of nature, culminating in the recent discovery of completely unified field theories based on the superstring. These theories locate a single, universal, unified field of intelligence at the basis of all forms and phenomena in the universe.[4]

THE NEW ENLIGHTENMENT

TM training begins with a free lecture on the "benefits" of the technique, complete with scientific-looking charts of the body's physical response. People who decide to take the plunge are charged a fee, currently $2,500, for initiation during which they are assigned their own personal mantra and have their first meditation. Then they attend a series of three follow-up lectures where they learn about the proper practice of the technique and listen to tapes of Maharishi. At one time it was claimed that TM would lead to "enlightenment" within five to eight years, but this claim has apparently been dropped.

Exactly what constituted the promised enlightenment seemed to change from year to year, as Maharishi meditated privately on the question at the beginning of each year. For example, 1970 was the Year of Scientific Research, 1973 the Year of the World Plan, 1978 the Year of Invincibility to Every Nation, and so on. None of it was very enlightening unless, of course, you became an adept.

Geoff noticed that there were two types of people attracted to the movement. One was composed of counterculture hippies who assumed that the age of enlightenment would be "a global love-in with free sitar music." The other group of Maharishi's followers assumed that their service to the movement would be rewarded by a mansion with a staff of servants, a position of leadership in the coming world government, and the gratitude of all humanity.[5]

By 1980, responding to the public's newly discovered thirst for medical alternatives, Maharishi Ayurveda Products International was launched, selling everything from herbal toothpaste to "mind-body beverages." Other Maharishi enterprises included Jyotish Gems, where you could buy emeralds and rubies your astrologer (provided by the movement, of course) prescribed to ward of the bad effects of planets predicted in your horoscope.

And of course there were all kinds of organic produce, such as Maharishi Organic Honey, that you could order off a Web site.

The biggest Maharishi enterprise, however, was an accredited university. The central campus purchased in Fairfield, Iowa, in 1974 was called Maharishi International University. In 1995 its name was changed to Maharishi University of Management.

Geoff Gilpin attended MIU, graduating in 1978. At first he had fun, feeling like "I was part of a family with all the love and support, but with a purpose, something bigger than yourself."[6] He was slim and healthy from the strict diet and lifestyle. Maharishi kept promising enlightenment around the corner, but it never came. Furthermore, the movement became increasingly authoritarian and conservative, with suits and ties required even when sitting around late at night doing homework—in case a *Time* magazine reporter should drop by. The movement also became involved in politics, which seemed to Geoff at odds with a life of meditation and spiritual evolution. So when he graduated, Geoff drifted from the movement to the more mundane life with a wife and a software documentation business. To this day he continues his meditation and while he is pretty sure the effects are purely materialistic and psychological, he is not quite ready to do away with the word "spiritual" in describing what he perceives as "unique and powerful results of yoga and meditation."[7]

GRAND UNIFICATION

At one time it was thought that Earth was a place separate from the heavens, with different physical laws. On Earth, bodies tended to fall down, while celestial bodies tended to move in circles. But then Newton, after being hit on the head with an apple falling from a tree (or so the story goes), realized that the force that pulled the apple down from the tree also pulled the moon toward Earth, keeping it from flying off into space. That is, gravity on Earth was the same force as gravity in the heavens. And so the two forces were unified.

Until the twentieth century only two other fundamental forces besides gravity were known: electricity and magnetism. In the mid-nineteenth century, Michael Faraday (d. 1867) showed that electricity and magnetism were the same phenomenon. In 1865 James Clerk Maxwell unified electricity and magnetism in a set of elegant equations we call *Maxwell's equations* of electromagnetism. From these he derived the existence of electromagnetic waves that traveled at the speed of light in a vacuum, strongly implying that light was an electromagnetic wave.

In the twentieth century two new fundamental forces were discovered that operated only at nuclear and subnuclear distances: the strong nuclear force, which is the glue that holds nuclei together, and the weak nuclear force, which is responsible for the decay of nuclei in which an electron and antineutrino are emitted. The latter provides the source of energy at the center of the sun.

In the 1970s the so-called standard model of elementary particles and forces was developed that has remained consistent with all the data as of this writing, although it is expected to be supplanted in a decade or so as new data come in from the Large Hadron Collider in Geneva, Switzerland, and other new experiments. An important part of the standard model is the unification of the electromagnetic and weak forces into a single *electroweak force.*

After the remarkable success of the standard model, theorists immediately set to work applying similar principles in an attempt to unify the electroweak and strong nuclear force in what was called a GUT, or *grand unified theory.* Without data to guide them, however, they had a wide choice of unification schemes. In such a situation, you try the simplest first, and that was done by Sheldon Glashow and Howard Georgi in 1974 and given the technical name *minimal SU(5).*[8]

PROTON DECAY

While Capra was trying to make physics look like Eastern mysticism, Maharishi was trying to make Eastern mysticism look like physics. As mentioned, Maharishi originally studied physics. His ear caught the catchy term "grand unification" and pretty soon flyers were appearing in which the cosmic field of consciousness, with which TM supposedly put you in contact, was associated with the grand unified field.

The problem was, minimal SU(5) was falsifiable. It made a very specific prediction that protons will decay with an average lifetime of about 10^{31} years. That's a pretty long time, but measurable with the technology of the 1980s. You just have to watch a lot of protons. In the early 1980s several experiments were mounted capable of detecting proton decay at this rate. The two most sensitive, placed deep underground to shield them against cosmic rays, were inside the Fairport Salt Mine near Cleveland, Ohio, and in the Mozumi zinc mine in Kamioka, Japan. They were designed primarily to detect the reaction $p \rightarrow e^+ \rightarrow \pi^0$, where e^+ is an *anti-electron* or *positron* and π^0 is a neutral *pi meson* or *pion.* This was the most likely proton decay channel according to the theory.

After several years of operation in which no decays were seen it became clear that protons did not decay at the predicted rate. The current best experimental lower limit on proton decay is 1.6×10^{33} years, two orders of magnitude higher than the prediction. This was published in 1998 from a second-generation experiment at Kamioka in which I collaborated called Super-Kamiokande.[9] In short, minimal SU(5) has been soundly falsified as a grand unification scheme.

MORE GUTS

The demise of minimal SU(5) did not cause GUTs to disappear from TM literature. In 1984 a respected physicist named John Hagelin joined Maharishi International University, establishing there a graduate program in theoretical physics and taking on leadership in many of the university's activities.

Hagelin received his PhD from Harvard in 1981, where he had worked under Georgi. In chapter 2 we saw that Hagelin was one of the contributors to the film *What the Bleep Do We Know!?* as well as a three-time candidate for president of the United States.

As I mentioned when I discussed minimal SU(5), the lack of any empirical guidance has left theorists with a wide choice of options for the next step beyond the standard model. The grand unified theories all attempt to bring together the electroweak and strong forces, which are treated independently in the standard model.

One particularly interesting GUT that appeared in the late 1980s was called *flipped SU(5)*. TM literature would have you think it was (1) primarily the work of Hagelin and (2) a highly successful GUT fulfilling Einstein's dream of a unified field theory. Here's what Hagelin's Web site says as of this writing:

> He is also responsible for the development of a highly successful Grand Unified Field Theory based on the Superstring. Dr. Hagelin is therefore at the pinnacle of achievement among the elite cadre of physicists who have fulfilled Einstein's dream of a "theory of everything" through their mathematical formulation of the Unified Field—the most advanced scientific knowledge of our time.[10]

The earliest reference to flipped SU(5) that I could find is a 1982 singly authored paper by Stephen Barr.[11] A 1984 paper lists three authors, not including Hagelin.[12] Hagelin is one of four coauthors of a 1987 paper.[13]

While flipped SU(5) was a promising theory, it was hardly the answer to Einstein's "dream" of a unified field theory. For one thing, as with all GUTs, gravity is not included and Einstein's dream was based on his hope that he could extend general relativity, which is his theory of gravity, to include the other forces. Today's unification attempts such as string theory and various proposals for quantum gravity seek a "theory of everything" (TOE) that includes all four forces.[14] It remains to be seen whether any of these attempts will succeed.

POSTSCRIPT

Maharishi Mahesh Yogi died at his home in Holland on February 5, 2008, at the estimated age of ninety-one.

NOTES

1. Ashvaghosha, *The Awakening of Faith*, trans. D. T. Suzuki (Chicago: Open Court, 1900), pp. 79, 86, as quoted in Fritjof Capra, *The Tao of Physics* (Boulder, CO: Shambhala, 1975), p. 277.

2. Geoff Gilpin, *The Maharishi Effect: A Personal Journey through the Movement That Transformed American Spirituality* (New York: Tarcher/Penguin, 2006).

3. Herbert Benson, *The Relaxation Response* (New York: HarperCollins, 2000); first published in 1975.

4. Yogic Flying Clubs for Students, http://www.yogicflyingclubs.org/foundation.html (accessed May 27, 2008).

5. Geoff Gilpin, unpublished.

6. *The Maharishi Effect*, p. 28.

7. Geoff Gilpin, private communication.

8. Howard Georgi and Sheldon Glashow, *Physical Review Letters* 32 (1974): 438.

9. The Super-Kamiokande Collaboration, "Search for Proton Decay via in a Large Water Cherenkov Detector," *Physical Review Letters* 81 (1998): 3319.

10. John Hagelin Web site, http://www.hagelin.org/about.html (accessed May 27, 2008). This quotation has tended to move around different Web sites.

11. Stephen Barr, *Physics Letters* B 112 (1982): 219.

12. J. Derendinger, J. Kim, and D. V. Nanopoulos, *Physics Letters* B 139 (1984): 170.

13. I. Antoniadis, J. Ellis, J. Hagelin, and D. V. Nanopoulos, *Physics Letters* B 194 (1987): 231.

14. Lee Smolin, *Three Roads to Quantum Gravity* (New York: Basic Books, 2001).

5

SPACE, TIME, AND MATTER

As children in blank darkness tremble and start at everything, so we in broad daylight are oppressed at times by fears as baseless as those horrors which children imagine coming to them in the dark. The dread and darkness of the mind cannot be dispelled by the sunbeams, the shining shafts of day, but only by an understanding of the outward form and inner workings of nature.

—Lucretius, *The Nature of Things*

MATERIALISM AND NATURALISM

Materialism is the doctrine that the universe is composed of a single substance called *matter* and nothing else. The answer materialism gives to the question "Is anything out there?" is: Nothing is out there except matter.

Materialism is usually equated with *naturalism*, which is probably best described as the doctrine that denies the existence of gods or other "spirits." These entities are then, by definition, *supernatural.*

One might imagine a universe with no gods or spirits but still including stuff other than matter that we can still reasonably label as natural. By "stuff" here I refer to concrete, objective substances and exclude various abstractions like human thoughts and emotions, words, or mathematical equations. Think of the stuff as still being there even if humanity did not exist. For example, we might find substances that did not behave like matter but yet obeyed some identifiable "laws of nature" separate from the laws of nature we associate with matter. Since we have never seen any sign of such stuff, let's not bring up that possibility until the data require it. For our purposes we will equate any immaterial stuff with supernatural stuff. The *materialist/naturalist view* then is that there is only matter. The *spiritualist/supernaturalist view* then is that there is matter and spirit. I will not take seriously the *idealist view* that there is only spirit. Samuel Johnson quickly refuted that by kicking a rock. The rock kicked back.

Often you will hear that science only deals with the natural. This is a position that is held by many scientists and is in fact the official doctrine of the National Academy of Sciences, the most prestigious scientific institution in the United States:

> Science is a way of knowing about the natural world. It is limited to explaining the natural world through natural causes. Science can say nothing about the supernatural. Whether God exists or not is a question about which science is neutral.[1]

This has led to the charge from theists that science is dogmatically opposed to the supernatural or the immaterial. This charge is understandable given the academy's position, but nevertheless it is unfair. We can rest assured that if any evidence were found for a world beyond matter scientists worldwide will be only too delighted to accept private or governmental funding to research the phenomenon. And the Academy won't stop them.

Furthermore, the NAS is dead wrong factually since reputable scientists in prestigious institutions such as Harvard University, Duke University, and the Mayo Clinic have done research on the efficacy of intercessory prayer, which surely is of supernatural significance.[2] So far the evidence is negative.

In fact, the purely material universe is nothing more than a working assumption, a scientific model that scientists have proposed to describe the data. So far this is all that is needed. If and when this model proves to be insufficient, then science will have to consider other possibilities.

ORIGINS OF MATERIALISM

Most of human history is the story of people dominated by shamans and priests who demanded worship, sacrifice, and strict obedience to unseen gods and spirits—as the shamans and priests interpreted the desires of gods and spirits. Materialism has ancient origins. In the sixth century BCE, Thales, of Miletus (ca. 546 BCE)—a Greek colony on the coast of what is now Turkey—proposed that everything was made of water. In the following century, the Greeks Leucippus (ca. 440 BCE) and Democritus (ca. 400 BCE) imagined that matter was composed of elementary "atoms" that could not be broken down further. Similar ideas are said to have appeared in India around the same time.

Two centuries later, Epicurus (d. 270 BCE) introduced a philosophical school of thought in which no gods existed and the universe was composed of atoms moving in empty space. He emphasized living a happy, self-sufficient life with no expectation of an afterlife. Although self-indulgence is often associated with the epicurean lifestyle, Epicurus emphasized personal responsibility and morality.

The philosophy of Epicurus inspired one of the great ancient poems, *De Rerum Natura*, or *The Nature of Things*, by the Roman Lucretius (d. 55 BCE), which is dated in the first century BCE. This poem was the only major work on classical materialism to survive antiquity intact.

Any voices of materialism that may have remained inside Christendom in the Dark Ages were suppressed by the Church until reappearing in the seventeenth century in the writings of Francis Bacon and Pierre Gassendi. In 1770 Paul d'Holbach published a monumental work, *The System of Nature*, which emphasized a materialistic, atheistic worldview in obvious contradiction to prevailing beliefs. In 1884 another influential manuscript appeared, *Force and Matter: Principles of the Natural Order of the Universe*, by Ludwig Buechner. This laid the groundwork for the modern materialistic view of nature that forms the underlying bedrock of science and much of secular thinking today.[3]

AUGUSTINE AND TIME

The materialist model of the universe is one of particles of matter moving about in space and time. Let us begin by taking a look at time.

In the fourth century, the great theologian Augustine of Hippo (d. 430) asked what God was doing before he made heaven and Earth. If he did nothing, why didn't he continue to do nothing? If he did something, performed some act, then that could not be part of true eternity since whatever the result of the act, it did not previously exist. How is it possible for an eternal God to do anything not eternal? Augustine jokes that God was busy preparing hell for those who "pry into mysteries." But he concludes more seriously that God is timeless, that time itself is not part of ultimate reality. Rather, time is subjective, existing only in the human mind, created by God. God thus lives in a different world than humans—a timeless one. According to Augustine, time as we humans know it was created by God "when" he created the universe.

As for the world of time in which humans live, Augustine gave no good explanation for why God put the idea of time in our heads in the first place. Living well before humans had accurate clocks, he was not sure how to measure time, what it is when you say one period of time is longer than another.

However, Augustine's insight that time exists only in the human mind was right on the mark.

TIME: A HUMAN INVENTION

No human observation would seem to be so ubiquitous as the passage of time. Physicists and philosophers long after Augustine have never come up with a satisfactory explanation of time. Usually they try to describe time in terms of the concept of change, but how do you define change without having a notion of time to begin with?

In his younger days before he became more metaphysical, Einstein gave a definition of time that remains to this day the best we can do: *Time is what you read on a clock.* This is an example of what we call an *operational* definition. All measurable quantities in physics are defined by how they are measured. Temperature is what you measure on a thermometer. Electric current is what you measure on an ammeter. Time is what you measure on a clock.

Like all the quantities of physics, time is a human invention. Of course, it is an invention used to describe observed phenomena in the external world, but it is a mistake to assume that what physics defines as time is identical to some metaphysical river that flows through the universe. Keep that in mind and you will have less difficulty accepting those modern physics notions of time that defy common sense.

In physical models time is usually represented by a real number. However, in cosmology you may read about "imaginary time," which is simply a mathematical construct whereby the measured time is multiplied by –1 in order to provide for more convenient calculations. We will run into it in the last chapter when we discuss two scenarios for the natural formation of the universe from "nothing."

Recall that Augustine had the deep insight that God, assuming he exists, operates outside of time. This is not hard to understand when we realize time is a human invention and so need not apply to God.

Throughout most of history, the passage of time was registered by familiar regularities such as day and night and the phases of the moon, or more accurately by the apparent motions of certain stars. The second was defined by the ancient Babylonians to be 1/84,600 of a day. Our calendars are still based on *astronomical time* using the Gregorian calendar, introduced in 1582, in which the year is defined as 365.2425 days. More accurately, 1 year = 365.242199 days, from modern estimates.

Until the scientific revolution and the ages of exploration and industrialization that followed, most people had no need for accurate clocks. Farmers and fishermen measured time in relation to familiar processes in the cycle of work and domestic chores. Labor took place in the natural period from dawn to dusk. The sundial was widely used to tell time during the day. The great advance in the accuracy of household clocks came about in the mid-seventeenth century with the application of the pendulum, which had been introduced into scientific experiments by Galileo Galilei (d. 1642) in 1602. English clock- and watch-making became dominant in 1680 and remained so until competition from the French and Swiss caught up about a century later.

In 1759 John Harrison (d. 1776), seeking to win an English Parliamentary prize, produced a clock, or *chronometer*, that could keep exact Greenwich Mean Time at sea, enabling mariners to determine their longitude on the globe and making accurate marine navigation far from land possible for the first time.[4]

Today the primary time standard is provided by averaging the outputs of a bevy of Cesium Fountain atomic clocks at the National Institute for Standards and Technology laboratory in Boulder, Colorado, near where I live, which will not gain or lose a second in more than 60 million years.

With the rise of science, the standard unit of time, the second, has undergone several redefinitions to make it more useful in the laboratory. The most recent change occurred in 1967 when the second was redefined by international agreement as the duration of 9,192,631,770 periods of the radiation corresponding to the transition between the two hyperfine energy levels of the ground state of the $Cesium^{133}$ atom at rest at absolute zero. If you think of an oscillating electromagnetic wave being emitted by the atom and using each peak in the wave to move the clock one tick, then 9,192,631,770 ticks would correspond to one second. In other words, the time between ticks is 1/9,192,631,770 of a second, or about 0.11 nanoseconds, where a nanosecond is a billionth of a second.

In short, time, as we use it in both science and everyday life, is simply the number of ticks on a clock.

The minute remains 60 seconds, the hour remains 60 minutes, and the day remains 24 hours, following ancient traditions. The day is still taken to be 84,600 seconds, as in Babylonia. Our calendars need to be corrected occasionally to keep them in harmony with the seasons because of the lack of complete synchronization between atomic time and the motions of astronomical bodies.

THE SMALLEST TIME INTERVAL

Time intervals have been measured as small as 10^{-16} second as of this writing. However, we cannot continue to divide time into smaller and smaller units. Because of both relativity and quantum mechanics, which we will describe later, the smallest operationally definable time interval is the *Planck time*, 6.4×10^{-44} second. This means that, fundamentally, time is an integer number of Planck units. It is (by definition) *discrete*, occurring in jumps, rather than continuous.

The discreteness or granularity of time is so small that it plays no role in even the most precise measurements of contemporary physics. As a result, physicists usually represent time in their equations with the real number t, which is assumed to be continuous. However, it is important to recognize that assuming time is continuous is an approximation. In fact, it is discrete.

TIME: LIMITLESS BUT NOT INFINITE

In figure 5.1 a discrete time axis is shown in which each marker represents one step in Planck units. Suppose we define "now" as $t = 0$. Then we can count steps in the direction that we call the future: +1, +2, +3 and so on. Nothing that we know about the universe from cosmology and physics requires that we must stop counting, terminating the sequence at some point in the future. Of course we will all be dead and Earth extinct some time in the future; the universe may eventually be devoid of life in any form, but any clock that is still out there, such as a spinning neutrino, will keep ticking. From this we infer that time is limitless.

Similarly, we can start now and count back −1, −2, −3, and so on into the past. Despite the claims of theologians from Augustine onward that the universe had a beginning in time, nothing we know about the universe from cosmology and physics requires that we stop counting, terminating the sequence sometime in the past. The universe does not appear to have any time limit in the past as well as in the future. Most likely it always was and always will be—just as it appeared to science before the discovery of the big bang.

Now to say that something is limitless is not the same as saying it is infinite. The term *infinity* is often used loosely, even by physicists, to refer to a very large number. However, from the work of the nineteenth-century

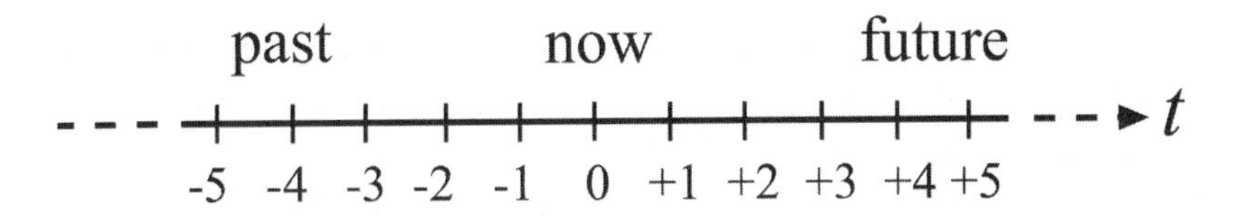

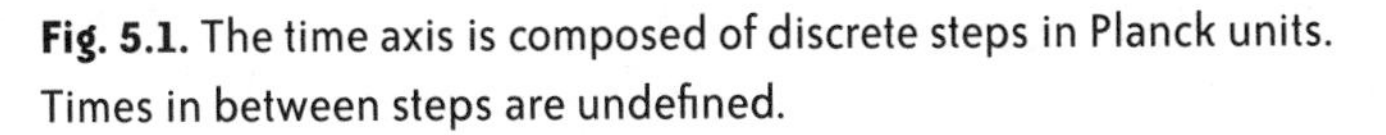
Fig. 5.1. The time axis is composed of discrete steps in Planck units. Times in between steps are undefined.

mathematician Georg Cantor, the set of real numbers we use to count the ticks on a clock form an infinite set, but none of the members of the set are themselves infinite.

THE ORIGIN OF TIME

Any point in time can be taken to represent the origin of our time axis, the point we call $t = 0$. In figure 5.1 I chose that origin to be "now." In the West we count years starting four years after the supposed birth of Christ. We count forward in years we label as CE, the Common Era (previously AD) and backward as BCE, Before the Common Era (previously BC). Obviously that is an arbitrary choice not followed on other calendars such as those of Chinese, Jews, and Muslims. Some confusion arises since the first year of the Common Era is not year zero but 1 CE, while the preceding year is 1 BCE. Astronomers correct this in their own calendars by calling 1 CE year 1, 1 BCE year zero, 2 BCE is year –1, and so on.

When we time a race with a stopwatch, we reset the time to zero so we can read the race times directly off the watch. When we do so we are implicitly assuming a basic principle about time: *time intervals do not depend on when those intervals are measured.*

Now, we can imagine a world in which this was not the case. Suppose clocks ran differently at different times of day. That could happen if the standard clock we used to define time was, say, my heartbeats. Each morning I try to get some exercise such as playing tennis or taking a vigorous walk. After lunch I usually nap in my chair. So if you were timing a race by my heartbeats, a morning race would generally lead to longer race times than races after lunch.

Obviously we have avoided such complications by using objective means of measuring time, with the current definition in terms of atomic vibrations being an improvement over heartbeats or even over previous astronomical measures where, for example, using Earth's complicated orbital dynamics would lead to small but similar discrepancies.

So, with time defined as objectively as possible we find that we can describe phenomena such as the motion of a runner or elementary particle in a way that does not depend on the origin of time. Putting it another way, the universe does not seem to single out any special moment in time.

But what about the beginning of the universe? Wasn't that a special moment in time? Wasn't it the beginning of time?

WHAT ABOUT THE BIG BANG?

Until the twentieth century, science had no evidence that our universe was of finite age. The stars and planets year by year repeated their motions through the skies to great precision. Rare cosmic events such as eclipses and comets were shown to also repeat in a predictable way. In the short period that humans have been making scientific observations of Earth and sky, these predictable cycles did not seem to change. For all anyone knew, the universe always was.

The discovery that the universe is expanding changed all that. Cosmologists are now certain that our universe appeared some 13.7 billion years ago in an explosion called the *big bang*. Many theologians and even one pope have publicly asserted that this discovery confirms the existence of a creator God. This claim appears frequently in popular Christian literature. For example, the well-known conservative author and political commentator Dinesh D'Souza writes in his 2007 book, *What's So Great about Christianity?*

> In a stunning confirmation of the book of Genesis, modern scientists have discovered that the universe was created in a primordial explosion of energy and light. Not only did the universe have a beginning in space and time, but the origin of the universe was also a beginning *for* space and time.[5]

Now, in fact the Genesis story of creation bears no resemblance to that described in big bang cosmology and, indeed, is in deep conflict with it. It has Earth created before the sun, moon, and stars. Actually, Earth formed eight billion years after the first stars. The Bible can hardly be credited with predicting the expanding big bang when it describes the universe as a firmament with Earth fixed and immobile at its center.

Every culture has creation myths of one kind or another, none more than superficially resembling the big bang. None can claim that their particular beliefs have been scientifically confirmed by the big bang. The Chinese myth actually comes closer than Genesis, starting in chaos and exploding into an expanding universe.

Nevertheless, the assertion is made that religion was ahead of science in conceiving of a universe of finite age, one that had a beginning. What is more, the physics of the big bang seemed to confirm Augustine's intuition that time itself began when the universe was created. This idea has been heavily promoted in the last few decades by the Christian apologist and public debater William Lane Craig. He calls it the Kaläm cosmological argument.[6] As he phrases it,

Kaläm Cosmological Argument

1. Whatever begins to exist has a cause.
2. The universe began to exist.
3. Therefore, the universe has a cause.

That cause, of course, is the "first cause" of Aristotle and Aquinas that people call God.

Craig bases his claim that time began with the big bang on the notion that the universe began as *singularity*—an infinitesimal region in space in which the mass and energy densities are infinite. This was believed to be the case. In 1970 cosmologist Stephen Hawking and mathematician Roger Penrose used Einstein's general theory of relativity to "prove" that our universe began with a singularity.[7]

However, over twenty years ago Hawking and Penrose realized that no such singularity marked the beginning of our universe. Indeed, Hawking explicitly says so in his phenomenal 1988 best seller, *A Brief History of Time*.[8] The original proof of Hawking and Penrose was not in error as far as it went. General relativity does imply the singularity. However, the authors now admit that because of quantum mechanics, general relativity does not apply below a minimum distance equal to the Planck length and below a minimum time interval equal to the Planck time. In fact, as I argued above, these are the smallest definable time and distance intervals so the universe could never have been an infinitesimal point. In short, time (or space) need not have begun with the big bang. As we will see in chapter 16, modern cosmological scenarios call for a universe prior to ours and, very likely, many more as well.

So, the universe need not have had a beginning, refuting Craig's Kaläm argument. But even if there was a beginning, it need not have had a cause. In his book, D'Souza ridicules me for making such a suggestion: "Physicist Victor Stenger says the universe may be 'uncaused' and may have 'emerged from nothing.' He quotes philosopher David Hume as saying, 'I have never asserted so absurd a proposition as that anything might rise without cause.'"

Well, Hume can be excused for not knowing about quantum mechanics. But D'Souza has no excuse for either not knowing or deliberately hiding the fact that quantum phenomena such as atomic transitions and nuclear disintegrations occur spontaneously without cause. Similarly, Craig has no excuse for continuing to use the singularity claim two decades after it was withdrawn by its authors.[9]

SPACE: ANOTHER HUMAN INVENTION

We have seen that, at least as used in physics, time is defined operationally as what is measured on a clock. While any clock, such as my heartbeats, could be used, this would make time intervals depend on my personal activity. Even a clock based on Earth's rotation would require periodic adjustments because of changing tidal interactions between Earth and the other objects in the solar system. So we have defined time in terms of what is read on a cesium atomic clock. Now let us operationally define *space.*

Recall that our clocks measure only time intervals. No "absolute time" can be identified, measured from some special moment. Similarly, we can only specify a spatial interval between two points, which we call *distance* or *length.*

The familiar units of length still used in America and one or two other countries are defined in the *English system.* An inch is about the thickness of the thumb of a grown male; a foot is about the length of his foot; the yard is about the length of his stride. This is not very objective, but these units are more accurately defined today in terms of the *meter* of the *metric system,* now the standard in most countries as well as in science.

In 1793 the meter was introduced as 1/10,000,000 of the distance from the pole to the equator. Since then it has gone through a series of increasingly precise definitions, from the length of a platinum-iridium bar stored under carefully controlled conditions in Paris to a certain number of wavelengths of the electromagnetic radiation from the krypton atom.

In 1905, Einstein introduced his *special theory of relativity*, profoundly revising our notions of space and time. The primary postulate of relativity is that the speed of light in a vacuum, a quantity conventionally referred to as *c*, is a constant that does not depend on the motion of the source of light or its observer. Thousands of scientific observations in the century since have confirmed the validity of Einstein's postulate.

In 1983, by international agreement, Einstein's postulate was incorporated into the definition of the meter, which was then defined to be the distance traveled by light in a vacuum during 1/299,792,458 of a second. As discussed earlier, the second is defined as a certain number of vibrations of the cesium atom.

This latest definition of the meter has a profound consequence that is not widely recognized even among physicists. Since 1983, distance is no longer treated as a quantity that is independent of time. In fact, as we see from the definition of the meter above, distance is now officially defined in terms of time. Distance is the time it takes light to travel between two points in a vacuum. Of course, in practice we still use meter sticks and other means to measure distance, but in principle these must be calibrated against an atomic clock.

A further implication of the definition of the second and the meter is that the quantity *c* called "the speed of light in a vacuum" is simply an arbitrary conversion factor. If you measure time in seconds and distance in meters, then *c* is *by definition* 299,792,458 meters per second. If you measure time in years and distance in light-years, $c = 1$ light-year per year, since the light-year is defined as the distance traveled by light in one year. When light travels through a medium, however, its speed is given by *c* divided by the index of refraction of the medium. Since no perfect vacuum exists in the universe, light can generally be found moving at a speed other than *c*, although the difference is very small in a near vacuum such as that of outer space.

THE SMALLEST SPACE INTERVAL

We have seen that the smallest operationally defined time interval is the Planck time, 6.4×10^{-44} second. As long as we stick to the operational definition of time as what you read on a clock, then this becomes the smallest measurable time interval.

From the previous section, the operational definition of a space interval, what we call distance or length, is also what you measure on a clock as light moves in a vacuum from one end of the space interval to another. If the smallest measurable time interval is the Planck time, it follows that the smallest interval in space is the speed of light in a vacuum $c = 299{,}792{,}458$ meters per second times the Planck time. This distance, 1.9×10^{-35} meter, is called the *Planck length.*

And, just as a time interval can fundamentally be viewed as an integer number of Planck times, so a space interval can fundamentally be viewed as an integer number of Planck lengths. As with time, space is discrete. And, as with time, the fact that distance is usually thought of as a continuous variable is an approximation that is fine for most purposes but breaks down at the Planck scale. In other words, there exists no "space-time continuum" in any proper model describing physical events. We can get away with assuming a continuum for most applications but we should be warned not to draw any universal or metaphysical conclusions from such models.

LIMITLESS BUT NOT INFINITE SPACE

In figure 5.1 a discrete time axis is shown in which each marker represents one step in Planck units. We can do the same for space, where we define "here" as $x = 0$. Then we can count steps in the positive direction on the x-axis: +1, +2, +3, and so on. Nothing that we know about the universe from cosmology and physics requires that we must stop counting, terminating the sequence at some distant place. Obviously we can do the same along the negative x-axis: −1, −2, −3, and so on. The universe has no known boundary in either space or time.

As was the case for time, it is technically incorrect to say the universe is "infinite" in size. It simply is without end in any direction. And, just as there is no special moment in time, there is no special position in space—no center of the universe. This was a recognition a long time in coming to the human race.

MATTER

Next, let us discuss matter. A very simple definition of matter is *anything that kicks back when you kick it.* That is, matter has the property that physicists call *inertia.* One measure of a body's inertia is its *mass.* The greater a body's mass, the harder it is to get moving. It is also harder to stop once it is moving. Or, more generally, the more massive a body the harder it is to change its motion. This is common experience.

The inertial properties of a body are also described by the *linear momentum*, which Newton identified as the "quantity of motion." Linear momentum is a *vector*, that is, it has both *magnitude* and *direction.* It is closely related to the *velocity vector*, whose magnitude is the *speed* of a body—what you read on a speedometer. The direction of velocity tells you where the body is headed, as measured, for example, with a compass.

For speeds low compared to the speed of light, the linear momentum of a body is the product of its mass and velocity. A more complicated formula is needed at speeds near the speed of light. The direction of the linear momentum vector is in all cases the same as the direction of the velocity vector.

If you have an isolated system of bodies and add up their linear momenta vectorially, the bodies can exchange linear momenta by interacting with one another, as long as the *total* linear momentum does not change.

Law of Conservation of Linear Momentum

The total linear momentum of an isolated system of bodies is conserved.[10]

An example of conservation of linear momentum is the recoil you experience when you fire a gun. While the bullet has low mass, it has appreciable momentum by virtue of its high speed. That must be balanced by a recoil of your body, which is at lower speed by virtue of your greater mass.

The linear momentum is the quantity of linear motion, that is, motion in a straight line. The *angular momentum* of a body is the quantity of rotational motion, that is, motion in a circle. A ball being twirled in a circle on a string has an angular momentum equal to its linear momentum multiplied by the radius of the circle.

If you have an isolated system of bodies and add up their angular momenta vectorially, the bodies can exchange angular momenta by interacting with one another, as long as the *total* angular momentum does not change.

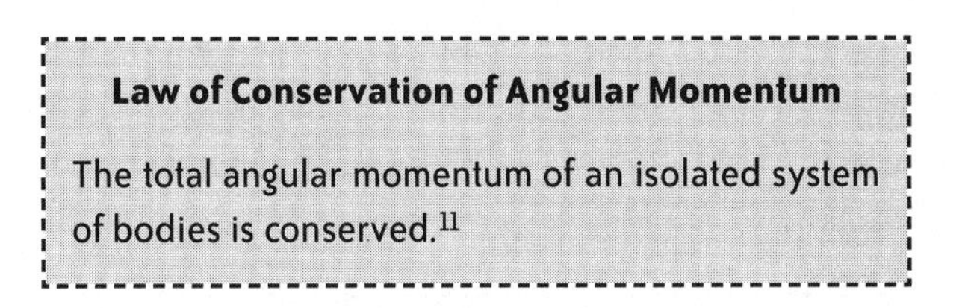

Law of Conservation of Angular Momentum

The total angular momentum of an isolated system of bodies is conserved.[11]

Angular momentum conservation is what keeps a moving bike from falling over.

Often physicists drop the "linear" from linear momentum. I will adopt that convention, so from now on when I mention "momentum," it will refer to linear momentum.

The final measure of a body's inertia is its *energy*. A moving body has an energy of motion called *kinetic energy*. A body, moving or at rest, can also have stored energy called *potential energy*. For example, a rock held above your head has potential energy that converts into kinetic energy when you release it and it falls to Earth.

In 1905 Einstein showed that a body also contains a *rest energy* E_0, equal to its mass m multiplied by the speed of light c squared—what all writers call "Einstein's famous equation," $E_0 = mc^2$. As with the rock example, energy can change its type and, because of $E_0 = mc^2$, mass can be converted to energy and, inversely, energy can be converted to mass. The total energy E of a body is thus the sum of its rest, kinetic, and potential energies.

If you have an isolated system of bodies and add up their energies, the bodies can exchange energy by interacting with one another, as long as the *total* energy does not change.

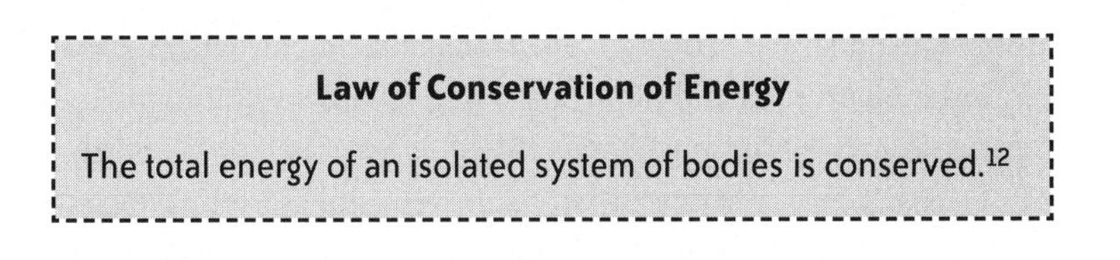

Law of Conservation of Energy

The total energy of an isolated system of bodies is conserved.[12]

For example, in chemical and nuclear reactions that produce heat, the kinetic energy that comes out is equal to the difference between the total rest energy for the initial reactants and that of the final reactants.

The law of conservation of energy is also known as the *first law of thermodynamics*.[13]

Material bodies are also affected by gravity, being pulled toward other bodies in proportion to the product of the masses of the bodies and in inverse proportion to the square of the distance between their centers.[14] This is called *Newton's law of gravity*.

Newton's Law of Gravity

The gravitational force between two bodies is proportional to the product of the masses of the bodies and inversely proportional to the square of the distance between their centers.

ATOMS AND PARTICLES

In common experience matter appears in three forms: solid, liquid, and gas. A sample of each looks smooth, but is in fact mostly empty space filled with a large number of tiny bits of matter called *molecules*. These molecules are composed of even smaller objects we call *atoms*. Atom is a bit of a misnomer, but in the nineteenth century it was thought that these were the elementary objects conjectured by Leucippus and Democritus, which were called atoms because they were assumed to be "uncuttable," not composed of simpler parts. Today's atoms are identified with the *chemical elements* of the Periodic Table. Elements cannot be broken down further by chemical reactions, since these have insufficient energy. They can, however, be "transmuted" by nuclear reactions.

Atoms have a substructure of electrons and nuclei, where the nuclei are composed of protons and neutrons. The protons and neutrons, in turn, are composed of two kinds of objects called *quarks*: the *up* (u) and the *down* (d). We see that a large variety of very short-lived material particles exists that is produced in high-energy collisions. Two additional "generations" of quarks and heavier versions of the electron are known to constitute these particles. While these two generations do not currently play an important role in the universe, they were very important in the early stages of the big bang.

I will provide evidence that light is also a form of matter, composed of particles we call *photons*. Photons have inertia and are also affected by gravity. Very low-energy photons left over from the big bang formed a highly uniform gas, cooled by the expansion of the universe to 2.7 degrees above absolute zero (Kelvin scale) that fills the universe. This is called the *cosmic microwave background.* While not a major contributor to the total energy of the universe, the number of photons in this background is a billion times greater than the number of atoms in all the galaxies.

MATTER IN THE UNIVERSE

Just three particles—the *u* and *d* quarks and the electron—are needed to describe atomic matter, which constitutes most of familiar matter, including all the matter in planets and stars including living organisms. By far most of the matter of the universe is invisible to both the naked eye and conventional telescopes. Visible matter—all the stars and galaxies that give off light—constitutes a mere 0.5 percent of the mass of the universe.

Only 3.5 percent of the remainder is composed of nonluminous atomic matter—dust, rocks, planets, and burned-out stars. We now know from its gravitational effects on the visible universe that 26 percent of the mass/energy of the universe resides in a yet unidentified form of invisible matter dubbed *dark matter.*[15] Cosmologists have ample evidence that dark matter cannot be the same kind of stuff as atomic matter, that is, it is not composed of quarks and electrons.

In 1998 it was discovered that the expansion of the universe was accelerating. This is currently explained as the action of some invisible stuff permeating the universe that actually has negative or repulsive gravity and is pushing all the stars and galaxies in the universe away from one another at an ever-increasing rate. This stuff is called *dark energy* and carries by far the most mass/energy of the universe, 70 percent. It is separate from dark matter, which exhibits familiar attractive gravity.

In his *general theory of relativity* published in 1916, Einstein showed that repulsive gravity was possible when a medium has a pressure that is sufficiently negative. He also introduced what is called the *cosmological constant*, which is the curvature of empty space and is equivalent to a field with negative pressure and repulsive gravity. Observations made since 1998 tend to favor the cosmological constant as the source of dark energy, but other possibilities remain.

Whatever the nature of dark matter and dark energy, they are clearly material and natural, having the properties of inertia and gravity that we associate with matter. Those seeking something supernatural out there need to look elsewhere than the dark matter and the dark energy.

THE SECOND LAW OF THERMODYNAMICS

Common experience tells us that time "flows" in one direction, from past to future. Would that it didn't and once in a while we could grow younger. The fact is that many observed phenomena such as aging, the deterioration of structures, and the wearing down of machinery do not ever reverse themselves. We never grow younger; structures don't spontaneously restore themselves; machines need to be regularly retooled.

These observations are codified in the *second law of thermodynamics*, which says that certain macroprocesses are irreversible. In its simplest form, proposed in the nineteenth century by Rudolf Clausius (d. 1888), the spontaneous flow of heat is always from a higher temperature body to a lower temperature one.

Second Law of Thermodynamics
(Clausius's simple form)

Heat cannot spontaneously flow from a material at lower temperature to a material at higher temperature.

This tells us why we have to pay for air-conditioning. We must input energy in order to force heat to go from a lower to higher temperature region. It also tells us why we cannot build a perpetual motion machine, where the heat generated by friction can be turned around to provide energy input to the machine. These and many other examples are allowed by the first law of thermodynamics (conservation of energy), so something else is happening to prevent them from operating in the reverse direction. That something is described by the second law of thermodynamics.

In 1865 Clausius introduced the concept of *entropy*, which is a measure of the disorder of a system. He reformulated the second law in terms of entropy (all the various forms of the second law can be shown to be equivalent):

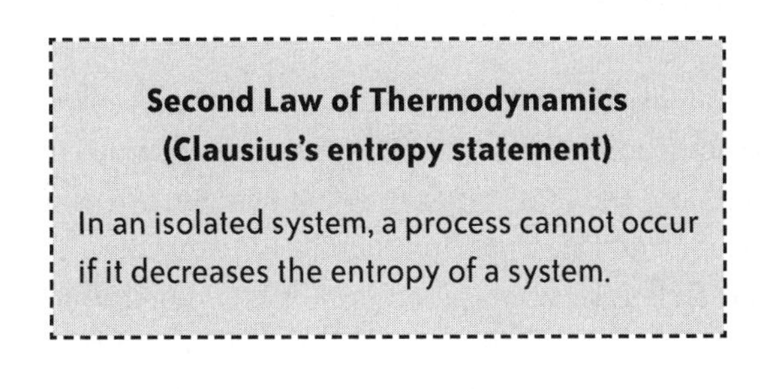

Second Law of Thermodynamics (Clausius's entropy statement)

In an isolated system, a process cannot occur if it decreases the entropy of a system.

Note the importance of the system being "isolated." When a system interacts with another system, its entropy can decrease, as long as the total entropy of whatever isolated system they are part, which may be the whole universe, stays constant or increases.

Later in the nineteenth century, Ludwig Boltzmann showed that the second law follows from the statistical behavior of a system of many particles in random motion. As these particles bounce off one another they tend toward a state of *equilibrium* in which the total energy of the system is shared among all the particles so that they each have the same average energy. A system in equilibrium is characterized by quantity called the *temperature* that is proportional to that average energy. It is also characterized by maximum entropy.

To see how the second law follows, bring two bodies of different temperatures together. The combined system will eventually reach equilibrium, but this will require the higher temperature body to cool off as its particles lose energy to the lower temperature body. Heat, which is just the flow of energy, will thus move from the higher to the lower temperature body and not the other way.

However, notice that the second "law" is not a principle that applies to individual particles. There is nothing stopping a particle in either body gaining or losing energy in any given collision. The phenomenon is just a probabilistic one that will happen only for a large number of particles and describes the system as a whole, not any of its parts.

The second law and other properties of thermodynamic systems are examples of "emergent" phenomena, which will be discussed in detail in chapter 10.

ENTROPY AND INFORMATION

In the late 1940s, as electronic communication mushroomed, a new technical field arose called *information theory* that studied ways to optimize the flow of wanted data and minimize the effect of noise. A definition of information was needed and this was provided in 1948 by Claude Shannon, who worked at the Bell telephone laboratory.[16] His definition turned out to equal within a constant factor the definition of entropy in statistical mechanics provided by Boltzmann. That constant is negative, so an increase in information is associated with a decrease in entropy. For this reason information is sometimes referred to as *negentropy*.

When a system is highly ordered, we think of it as having low entropy or containing a high level of information. In his book *Intelligent Design*,[17] the theologian William Dembski, who also has a degree in mathematics, proposed what he called the *law of conservation of information*. That is, he asserted that information is analogous to energy, momentum, electric charge, and other quantities of physics that do not change, in total, for an isolated system. Dembski used this law to "prove" that complex systems, systems of high information, cannot be made by natural processes from simpler systems of lower information. Thus, he attempted to argue, an intelligent designer is therefore required for all complex systems, in particular, living organisms.

Well, as I first showed in my book *Has Science Found God?* where I worked out the details, and as we can see from the above discussion about Shannon's definition of information (which Dembski claims to use, though he does so incorrectly), information and entropy are the same within a constant.

But entropy is not conserved. The second law of thermodynamics allows for the entropy of an isolated system to increase with time. It follows that the information in an isolated system can decrease with time.

Dembski's law of conservation of information is provably wrong. The universe can be an isolated system in which information or organization grows with time by means of purely natural processes within that system and with no need for any help from outside the system.

THE ARROW OF TIME

None of the basic equations of physics specifies a unique direction of time. Those equations work either way. Celestial mechanics can be used to predict exactly when and where a total eclipse of the sun will happen three thousand years from now. The same equations will tell you that a total eclipse occurred on the coast of Asia Minor on May 28, 585 BCE. Although the story is disputed, history records that Thales of Miletus predicted this eclipse and ended a war in the process. Whether or not he did, the event happened and physics confirms it with great precision!

According to basic physics, then, every process is reversible. For example, when you puncture a tire air will flow out, flattening it. But no basic law of physics forbids that a moment later air from the outside flows in and reinflates the tire. All that has to happen is that a sufficient number of outside air molecules move in the direction of the puncture. Since they are moving around randomly, there is some nonzero probability for this to occur. The problem is that probability is very small, so for all practical purposes the tire remains flat.

Another way to look at this is that the *arrow of time* is defined by the second law of thermodynamics, that is, by the direction in which the total entropy of the universe increases. It is the direction of most probable occurrences. Thus, the arrow of time of common experience is purely a statistical effect that results from the large number of particles in more-or-less random motion that constitutes material systems on the macroscale. This arrow is not a basic law of the universe. It is not present when the number of particles is few.

It is important to keep in mind, then, that the universe has no fundamental direction of time. Effects can precede causes and the whole idea of *creation*, which has a built-in assumption on the direction of time, needs to be rethought.

NOTES

1. National Academy of Sciences, *Teaching about Evolution and the Nature of Science* (Washington, DC: National Academy of Sciences, 1998), http://www.nap.edu/catalog/5787.html (accessed March 5, 2006).

2. M. W. Krucoff, S. W. Crater et al., "Music, Imagery, Touch, and Prayer as Adjuncts to Interventional Cardiac Care: The Monitoring and Actualization of Noetic Trainings (MANTRA) II Randomized Study," *Lancet* 366 (July 16, 2005): 211–17.

3. Richard C. Vitzthum, *Materialism: An Affirmative History and Definition* (Amherst, NY: Prometheus Books, 1995).

4. Dava Sobel, *Longitude: The True Story of a Lone Genius Who Solved the Greatest Scientific Problem of His Time* (New York: Walker Publishing Company, 1995).

5. Dinesh D'Souza, *What's So Great about Christianity?* (Washington, DC: Regenery, 2007).

6. William Lane Craig, *The Kalām Cosmological Argument*, Library of Philosophy and Religion (London: Macmillan, 1979).

7. Steven W. Hawking and Roger Penrose, "The Singularities of Gravitational Collapse and Cosmology," *Proceedings of the Royal Society of London*, series A, 314 (1970): 529–48.

8. Stephen W. Hawking, *A Brief History of Time: From the Big Bang to Black Holes* (New York: Bantam, 1988), p. 50.

9. William Lane Craig, "The Existence of God and the Beginning of the Universe," *Truth: A Journal of Modern Thought* 3 (1991): 85–96. Online at http://www.leaderu.com/truth/3truth11.html (accessed July 31, 2008).

10. Technically, the system need not be isolated, just the net force on it must be zero. *Force* is defined as the time rate of change of linear momentum.

11. Technically, the system need not be isolated, just the net torque on it must be zero. *Torque* is defined as the time rate of change of angular momentum.

12. Technically, the system need not be isolated, just the net energy into and out of the system must be zero.

13. Technically, the first laws say that the change in total energy of a system is equal to the heat in minus the work done by the system. Heat out is negative, as is the work done on a system.

14. Technically, the distance is between the *centers of gravity* of each body, which are the points where the gravitational force would be the same if all the mass were concentrated at that point.

15. These numbers can be expected to change slightly as measurements improve, but they define the ball park.

16. C. E. Shannon, "A Mathematical Theory of Communication," *Bell System Technical Journal* 27 (July 1948): 379–423; (October 1948): 623–25; Claude Shannon and Warren Weaver, *The Mathematical Theory of Communication* (Champaign: University of Illinois Press, 1949).

17. William Dembski, *Intelligent Design: The Bridge between Science and Theology* (InterVarsity Press, 1999).

6

THE GREAT PARADIGM SHIFT

The most important scientific revolutions all include, as their only common feature, the dethronement of human arrogance from one pedestal after another of previous convictions about our centrality in the cosmos.

—Stephen Jay Gould

THE COPERNICAN REVOLUTION

Humans in every culture of which I am aware have generally regarded themselves as special. In Europe, America, and the lands dominated by Islam, billions still hold the view that the universe was created—and continues to be ruled—by a personal God who made them in his image, gave them dominion over all life on Earth, and has reserved a glorious place for a select few (including them, of course) in eternity.

Until fairly recently, on the scale of human history, it was taken for granted that Earth rested immovably at the center of the physical world—exactly where it should if Earth's inhabitants were special. Indeed, the Bible seems to affirm this in several places. For example, Psalm 93:1, Psalm 96:10, and Chronicles 16:30 state that "the world is firmly established, it cannot be moved."

In 1543 Nicolaus Copernicus (d. 1543) triggered the modern scientific revolution with the publication of *De Revolutionibus Orbium Coelestium* (On the Revolutions of the Celestial Spheres) in which he proposed that Earth orbits the sun. A myth that has become attached to this story tells how he looked beyond the vanity of human self-centeredness and removed Earth from the center of the universe. Although Copernicus placed the sun at the center of the universe, astronomers soon realized that the sun is just another star, and a rather undistinguished one at that. That is, no point in space can be identified as special.

In fact, the belief that Earth is the center of the universe may not have had much to do at all with human self-centeredness.[1] Ancient people did not look at the center of the Earth as a desirable place. After all, it was the location of hell, while heaven was way out beyond the stars. But the Bible is pretty clear that Earth is absolutely fixed in space and that the sun, moon, and stars move with respect to Earth.

Aristotle (d. 322 BCE) had made the empirical argument that both Earth and the universe seem to have the same center, since heavy objects fall toward Earth and fire travels upward while celestial bodies seem to rotate about Earth.[2] This was still the scientific view at the time of Copernicus, with astronomers using the complicated geocentric system of Ptolemy (c. 85–165) to compute the positions of planets on the celestial sphere.

Copernicus's heliocentric system had been in circulation since 1515 and while other astronomers respected his ideas, they did not like the counterintuitive and apparently counterempirical notion of a moving Earth. If we are moving, why don't we notice it? Besides, the original version of Copernicus's model was no more accurate and only marginally less cumbersome than Ptolemy's, still containing some thirty epicycles—circles within circles. The Copernican picture did have a certain aesthetic appeal, however, offering a natural explanation for the zigzag motion of the planets as observed from Earth.

Eventually the Copernican model proved more successful, thanks to the careful observations of Tycho Brahe (d. 1601), which corrected a number of previous errors in astronomical data, and the realization by Johannes Kepler (d. 1630) that the planetary orbits were not circles, but ellipses. This got rid of the epicycles and located the sun at the common focus of the planetary orbits. Kepler also discovered that a planet sweeps out equal areas in equal times as it moves about the sun, now called *Kepler's law*. Sixty years later when Isaac Newton (d. 1727) trivially derived Kepler's law from his laws of motion and gravity, the Copernican system finally became firmly established.

HELIOCENTRISM AND THE CHURCH

Copernicus worded his proposal carefully and it did not cause an immediate religious hullabaloo. That would happen fifty years later at the hands of a brasher scientist, Galileo Galilei (d. 1642). Indeed, over the intervening period the Roman Catholic Church had adopted the heliocentric model as a calculational tool that, with the improvements mentioned, led to more accurate predictions of the positions of astronomical bodies. This was important for the accurate dating of Easter and other festivals. The Church has a long history of interest and support for astronomical research, going back to the fourteenth century when Pope Gregory XII (d. 1417) instituted the calendar we still follow today.

The story of Galileo's trial by the Inquisition in 1633 for teaching the Copernican system is often presented as a classic example of religion and science coming into conflict. But the story is also part myth and part fact. Historians now largely agree that Galileo was not tried for teaching heliocentrism, but for disobeying a Church order.

In 1610 Galileo had published the results of his telescopic observations of the heavens, which, in his mind, provided empirical confirmation of the Copernican view. He initially drew some support from powerful Church leaders, but in 1616 he was instructed not to discuss heliocentrism as a fact until he had definitive physical proof.

This he claimed to have in 1632 with the publication of *Dialogue on the Two Chief World Systems*. In fact the proof presented there, which was based on the assumption that Earth's motion causes the tides, was wrong. The argument was convoluted and incorrectly predicted only one high tide a day. Galileo dismissed as "useless fiction" the proposal by Kepler that the moon caused the tides, which turned out to be correct.

The scientific community today severely and justly criticizes any scientist presenting a poorly formulated proof that disagrees with the data. In 1633 Galileo was tried for disobeying the order of the Church. He agreed to a plea bargain in which he would admit he had gone too far, but for still unknown reasons the Inquisition overruled the agreement and handed down a harsh sentence in which Galileo was forced to recant.

Galileo lived out the final nine years of his life under comfortable house arrest, technically forbidden from writing further on physics. Somehow, however, he did some of best work during that time, publishing the *Discourse on Two New Sciences* that basically invented *kinematics*, the description of the motion of bodies. Isaac Newton would take off from there.

In the meantime, the Church promoted research into the Earth's motion and actually ran experiments in the 1650s and 1660s that provided empirical support for the Copernican system. Catholicism does not rely on the Bible as its final authority, but rather the pope, who they claim received that authority directly from Christ in an unbroken chain starting with Peter. When something in the Bible disagrees with science, all the pope has to do to resolve the conflict is declare that the scientific interpretation is right and it was God's idea all along. Unfortunately, popes have not done that very often.

When the Reformation rejected the authority of the pope, the defectors had no place else to go but the Bible for a replacement authority. And Copernicus clearly conflicted with the Bible. Martin Luther (d. 1546) called Copernicus a "fool who wished to reverse the entire history of astronomy."[3] John Calvin (d. 1564) denounced those who "pervert the course of nature" by saying that "the sun does not move and that it is the earth that revolves and that it turns."[4]

Perhaps feeling the pressure from Luther and Calvin, the Catholic Church after about 1650 began to regard the Copernican model as devaluing humanity and banned its teaching. In the meantime, as the evidence supporting the model became overwhelming, Protestant churches ended their opposition. The Catholic Church, being more tradition bound, bureaucratically clung to its anti-Copernican stance for almost another two centuries, not removing its ban on Copernicus until 1822. His book remained on the forbidden list until 1835, and in 1992 Pope John Paul II lifted the edict of inquisition against Galileo, 359 years after his trial.

YET IT DOES MOVE

Getting back to the physics, there was another problem with the Copernican model that was recognized by his fellow astronomers and by

Galileo's church critics. It violated common sense. How could Earth be moving at some high speed around the sun, actually thirty kilometers per second, and we humans, sitting here on Earth, not notice it? The experience of a seventeenth-century man or woman riding in a carriage down some cobblestone street seemed to clearly demonstrate that one could tell the difference between moving and being at rest. So, we should sense Earth's motion, but we don't. Legend has it Galileo insisted, *Epur si muove*—"And yet it does move." He probably didn't say it, but it is a good line.

Galileo could not conclusively prove that Earth moves. But he did have good reasons for believing the Copernican scheme despite common sense. There was significant observational evidence that favored the Copernican view. Galileo saw four moons revolving around Jupiter, which meant they did not revolve around Earth. He noticed that, like Earth's moon, Venus had a full set of phases including a fully illuminated phase that could only occur if Venus was sometimes on the other side of the sun. This would not happen in the geocentric picture, where Venus orbits Earth inside the orbit of the sun, which, in turn, revolves around Earth. Further observations, such as the craters and mountains on the moon, convinced him of the incorrectness of the picture drawn by Aristotle in which astronomical objects were perfect, smooth spheres, as befits their heavenly status.

Galileo's dogged insistence that he was right, even though he could not offer incontrovertible proof, would characterize the behavior of other scientists who made major discoveries in later years. When someone has a significant insight, it will sometimes just seem "right" by virtue of its elegance and simplicity. Einstein had that feeling about his theory of general relativity and perhaps special relativity as well. Neither was derived from direct observation, although they proved to agree with observations—the necessary element for them to remain science and not be dismissed as mere fantasies. Einstein's ideas certainly clashed with authority and common sense. Yet, he too had the courage to carry them forward and examine the consequences, no matter how outlandish these consequences turned out to be. In this he was carrying on the tradition of Galileo.

MOTION IS RELATIVE

Galileo exhibited courage. But he still had to explain the fact that we earthlings do not sense the Earth's motion. His answer was an extremely subtle one that required a deeper look at the meaning of motion. What we sense when we bounce up and down in a carriage on a cobblestone street is not motion itself, but *changes* in motion. Galileo and his contemporaries did not have our modern experience of flying in jetliners. Inside the jetliner cabin we have no sense of the fact that we are hurtling through the air at hundreds of kilometers an hour. Our sense of motion comes only from the changes in speed or direction we experience during takeoff and landing, and the occasional air turbulence we encounter during the flight.

Galileo distinguished between two measures of motion: *velocity*, which is the rate of change of position, and *acceleration*, which is the rate of change of velocity.

And finally, the models of physics cannot depend on the velocity of an observer. They should be the same for someone sitting on his porch as for an air force fighter pilot zooming by in a supersonic jet.

The fact that we cannot sense velocity is now recognized as a great principle of physics that more than any other proved Aristotle's physics to be grossly incorrect.

The Principle of Galilean Relativity

There is no observable difference between being at rest and being in motion at constant velocity.

Another way to state the principle of Galilean relativity is:

The Principle of Galilean Relativity (second form)

There is no way for an observer inside a closed capsule to detect or measure the velocity of that capsule.

Indeed, it can even be said that whether or not Earth moves is a matter of one's point of view—what physicists call one's *frame of reference.* In the point of view of someone sipping lemonade on his porch, the Earth is actually at rest! But from the point of view of an astronaut on the moon, Earth is moving. So neither rest nor motion at constant velocity, what is called *uniform motion,* is a universal concept that holds independent of the point of view of the observer.

THE PRIME MOVER

Although not realized at the time, Galileo's notion of the relativity of motion was far more heretical than his teaching that Earth is not the center of the universe. The Church was influenced heavily by the teachings of Aristotle during Galileo's time. In his treatise on physics written around 350 BCE, Aristotle wrote:

> Since everything that is in motion must be moved by something, let us suppose there is a thing in motion which was moved by something else in motion, and that by something else, and so on. But this series cannot go on to infinity, so there must be some first mover.

In the thirteenth century, Thomas Aquinas used Aristotle's prime mover argument as the first of the five ways he claimed to prove the existence of God. The other four arguments were little more than variations on the same generic argument. Aquinas defined four sets of agents, one agent following the other in each set. These sets of agents are: (1) movers; (2) causes; (3) contingent (dependent) beings; (4) greater beings; (5) purposeful agents. Then, without proof, Aquinas asserted that each set must have a first element—a prime mover; a first cause; a necessary (noncontingent being); a greatest being; an intelligent designer. Each of these beings Aquinas identified as what everyone calls "God."

Aquinas's proofs became the Church's dogma. However, Galileo showed that motion does not require a mover. Since no force is required to produce uniform motion, no prime mover God is necessary. The other four proofs are similarly based on a first premise that cannot be justified.[5] Aquinas had no basis for assuming the necessity of a first cause, a necessary being, a greatest being, or an intelligent designer. Thus all five of his proofs are refuted.

As we will see, quantum events can occur spontaneously—without cause. Also, recall the discussion at the end of last chapter about the arrow of time. The whole idea of cause and effect assumes a direction of time. The concept works well on the macroscale of human experience, but it does not apply at the fundamental level our universe.

NEWTONIAN MECHANICS

In 1687 Isaac Newton published *Philosophiæ Naturalis Principia Mathematica* (*Mathematical Principles of Natural Philosophy*), now referred to simply as *Principia*, which many scholars say is the greatest work of science ever produced. Newton incorporated Galileo's principle of relativity in his three laws of motion that represent the foundation of the mechanics of particles:

Newton's Laws of Motion

A body at rest tends to stay at rest and a body in motion in a straight line at constant velocity tends to stay in motion in a straight line at constant velocity unless acted upon by a net external force.

The total force on a body is equal to the rate of change of the momentum of that body.

For every action there is an equal and opposite reaction.

The first law of motion is essentially the principle of Galilean relativity with the additional definition of force as the agent that can change the velocity of a body. In common experience, bodies in motion do not remain in motion because of the ubiquitous presence of frictional forces that act to slow those bodies unless counteracted by a motive force such as an engine. The second law of motion simply defines force as the rate of change of the "quantity of motion," or *momentum* of a body. Actually, we have seen that two types of momentum are used in physics: *linear momentum*, which applies to straight-line motion, and *angular momentum*, which applies to rotational motion. Linear momentum can be approximated as the product of the mass and the velocity of a body except at speeds near the speed of light. Angular momentum is the product of the *rotational inertia* and *rotational velocity* of a body.

The first law implies the momentum of a body will be constant, or "conserved" in the absence of any net external force. If we assume the mass of a body is constant, then its velocity will not change unless acted on by an outside force. The second law tells us exactly what amount of force is needed to change the momentum of a body a certain amount in a certain time. The third law is a consequence of conservation of momentum, as exemplified by the recoil you experience when firing a gun.

These three laws, along with Newton's law of gravity and the laws of electricity and magnetism that were discovered in the nineteenth century, provide the fundamental principles of what is called *Newtonian* or *classical* physics. While classical physics has sophisticated theories that describe the structure of solids, the motion of fluids, the propagation of waves, and the behavior of thermal systems, these theories can all be derived from the fundamental principles of classical mechanics. Only in the twentieth century, with the development of relativity and quantum mechanics, have new principles been added to the old ones.

Often one hears that modern physics showed that Newtonian physics was proven wrong by the twin twentieth-century revolutions of relativity and quantum mechanics. Nothing can be further from the truth. To this day Newtonian mechanics remains the foundation of physics and the natural sciences that are built upon physics. The principles and methods of Newtonian physics are still taught in today's physics classes and continue to have great practical value. The domain of Newtonian physics encompasses most of common experience. It is utilized by most engineers and others who apply physics in their professions.

We have flown to the moon on Newtonian mechanics. The organs of the human body, including the brain, run on Newtonian mechanics. Relativity and quantum mechanics extended Newtonian physics to new domains that were not accessible earlier and showed how Newton's mechanics must be modified in these domains.

In 1962, physicist and historian Thomas Kuhn introduced the term *paradigm shift* into the lexicon of everyday discourse when his highly influential *The Structure of Scientific Revolutions* made that rare transition from academic to popular literature. The word *paradigm* refers to a worldview that underlies the theories and methodologies of a particular scientific subject. A paradigm *shift* occurs when a fundamental change occurs in a paradigm.

Kuhn argued that science proceeds by a series of paradigm shifts rather than small gradual changes. While other scholars have challenged this assertion and he later softened his claims, everyone agrees that the greatest paradigm shift of all time occurred when Isaac Newton, standing on the shoulders of Nicolaus Copernicus and Galileo Galilei, established the primacy of observation over authority and started us on the road to our modern scientific world.

NOTES

1. Mano Singham, "The Copernican Myths," *Physics Today* (December 2007): 48–52.

2. Thomas Kuhn, *The Copernican Revolution: Planetary Astronomy in the Development of Western Thought* (Cambridge, MA: Harvard University Press, 1957).

3. Martin Luther, *TableTalk*, 1539, as quoted in Granville C. Henry, *Christianity and the Images of Science* (Macon, GA: Smith & Helwys, 1998), p. 22.

4. Edward Rosen, *Copernicus and His Successors* (London: Hambledon Press, 1996), p. 159.

5. Roy Jackson, *The God of Philosophy: An Introduction to the Philosophy of Religion* (N.p.: Philosophers Magazine, 2001), pp. 24–37.

7

DEISM AND DARWINISM

Fix reason firmly in her seat, and call to her tribunal every fact, every opinion. Question with boldness even the existence of a god; because, if there be one, he must more approve of the homage of reason than that of blindfolded fear.

—Thomas Jefferson[1]

THE CLOCKWORK UNIVERSE

Until the twentieth century, Newtonian mechanics appeared to provide the means, at least in principle, for predicting the motion of a body with unlimited precision. All you need to know is the mass of the body, its initial position and velocity, and the net force acting on it. Then the laws of motion allow you to calculate the position and the velocity of the body at any later (or earlier) time.

Newton insisted that he was demonstrating the working of divine providence in nature. However, his discoveries conflicted profoundly with traditional Christian teaching. If the motion of every body in the universe is fully determined by Newton's laws of motion and force, then there is nothing for God to do beyond getting it all going at the creation—no reason to step in to perform miracles or answer prayers. Even before Newton, philosophers had begun to view the universe as a vast machine. Now that picture seemed to be confirmed; according to Newton we live in a *clockwork universe* with everything predetermined.

THE ENLIGHTENMENT

In the seventeenth and eighteenth centuries science and rational thinking began to challenge superstition and appeals to religious authority. This short period in Western history is referred to as the *Age of the Enlightenment.*[2] On the heels of the Reformation and the resulting religious strife, many factors worked together to undermine the authority of Christianity whether that authority was expressed through the pope or the Bible. Disgusted at the slaughter carried out in the name of religion during the wars between Protestants and Catholics, intellectuals and clergy alike began to look for alternatives to the prevailing forms of faith.

This period simultaneously saw the founding by George Fox (d. 1691) of the Quakers, the Religious Society of Friends, who sought light within their own hearts and said religion should be based on personal experience rather than doctrine. John Wesley (d. 1791) founded Methodism on similar ideas and the emancipation of faith from reason. The new forms of religion became divided between those who attempted to rationalize and demystify faith and those who jettisoned reason altogether.[3]

The thrust of the Enlightenment scholars was to trust in reason. Philosophers Francis Bacon (d. 1626), Thomas Hobbes (d. 1679), and John Locke (d. 1704) sought to find natural explanations of the world. One aspect was an attempt to reduce Christianity to the elements that are rational. As a result, a new concept of god arose, a new theology called *deism* in which a perfect, all-knowing, all-powerful god creates the universe and its laws, and then leaves it alone to carry on by itself. Whatever purpose this god had in creating the universe, that purpose is already built in and inevitable since every event is already predetermined. Indeed, only if God were imperfect would he need to intervene to change the course of events.

The terms *deism* and *theism* have the same roots, both stemming from the words for god in Latin and Greek, respectively. But they have come to mean something different. Theism continues to refer to belief in (usually) a single God as both the creator of the universe and one who maintains a personal relationship with humanity. Deism has come to refer to a creator who does not further intervene in the universe nor interact with humankind.

Historians date the beginning of the deist movement to 1624 with the publication of an essay by Edward Herbert, Lord of Cherbury (d. 1648), titled "On Truth, as It Is Distinguished from Revelation, the Probable, the Possible, and the False." Cherbury proposed a "natural religion" based on reason rather than revelation. Over the next two centuries other notable literary figures expounded on deism, including Charles de Secondat, Baron de Montesquieu (d. 1689), John Toland (d. 1722), François-Marie Arouet, better known as Voltaire (d. 1778), and Denis Diderot (d. 1784).

The "Bible" of deism is generally regarded to be *Christianity as Old as the Creation; or, the Gospel a Republication of the Religion of Nature* by Mathew Tindal (d. 1733), which first appeared in 1730. According to Tindal, the "true" religion must be perfect since it was created by a perfect god. He found no such perfection, no reason or common sense in Christianity as it was practiced, with centuries of persecution, terror, and strife. He found greater morality in Confucius and Cicero than in the Christianity of history.

The real revelation, Tindal claimed, is nature itself; the real god is the god Newton revealed. That god is the designer of a marvelous world operating majestically according to invariable law. The real morality is reason in harmony with nature. This is the true Christianity "as old as creation."[4]

Most Americans are familiar with the opening words of the Declaration of Independence, which mention in the Preamble, "The Laws of Nature and Nature's God," and then begin the main text with, "We hold these truths to be self-evident, that all men are created equal, that they are endowed by the Creator with certain unalienable Rights." It is widely assumed that these words refer to the traditional Christian God. However, this is hardly likely since the author, Thomas Jefferson, was a deist and an outspoken critic of institutional Christianity. He even wrote his own version of the New Testament that was stripped of all supernatural elements and presented Jesus as simply a great moral teacher.

Other likely deists among the Founding Fathers include Benjamin Franklin, Thomas Paine, and James Madison. George Washington and John Adams also appear to have held deist views. Washington refused to be attended by a clergyman at his deathbed. He oversaw the Treaty of Tripoli that declares "the Government of the United States of America is not in any sense founded on the Christian religion." The treaty was ratified by the Senate and signed by President John Adams in 1797.

The fourth president, James Madison, was perhaps the most critical of Christianity, saying, "Religious bondage shackles and debilitates the mind and unfits it for every noble enterprise." It can reasonably be said that the United States was founded on deist, not Christian principles.

A central tenet of the Enlightenment was Newtonian determinism. Some thinkers, particularly in France, saw no need even for a deist god, and for the first time in the history of Christendom atheism became a respectable alternative.

NATURAL THEOLOGY

The main objection to the clockwork universe was the implication that humans, as mechanical bodies themselves, do not possess free will. This meant that we are not responsible for our actions. Not only did this contradict the central religious doctrines of sin and atonement, it posed real problems for secular society. If a person is not responsible for his acts, what basis is there for punishing or rewarding him for those acts? Besides, most people have the innate conviction that they possess the freedom to act self-consciously no matter what scientists or philosophers may say.

In *Émile ou l'éducation* by Jean-Jacques Rousseau (d. 1778), which appeared in 1762, the Vicar of Savoyard chastises philosophers to learn to recognize that something may be true even if they cannot understand it. Such is the case for the free-acting, immaterial mind, which the vicar argues is a fact immediately perceived in one's "inner light." Rousseau led the way out of Enlightenment deism and atheism, teaching a theology in which everything natural is good, and evil is humanity's doing. Although he is often associated with the idea, Rousseau did not introduce the notion of the "noble savage." They were still humans, capable of evil.

Rousseau also disputed what has become the primary tenet of science since Galileo, that objective observation is the only means by which we can obtain reliable information about the world. Rousseau reaffirmed the still common belief in an immaterial mind capable of reaching beyond the world of our senses.

So the Enlightenment did not bring about the demise of Christianity. In America, deism appealed only to the educated few and quickly died out, as religion became marked with emotional excess in the second phase of what is called the *Great Awakening.* Although founded as a secular republic, the United States became increasingly Christianized and viewed internally as God's favored nation.[5]

In England and elsewhere in Europe, Christianity began to develop its own brand of rational religion called *Natural Theology*, using the metaphor of mechanism and the wonders of science to extol the glory and the power of God—which, in any case, was Newton's own view. Furthermore, nature itself ostensibly offered proof of God's existence.

The English archdeacon William Paley articulated this view eloquently in his 1802 book, *Natural Theology.*[6] There he introduced the famous watchmaker analogy for God. Paley tells of walking on the heath and finding a stone and a watch. While the stone is easily viewed as an object formed by natural forces, the same is certainly not true of the watch, which is clearly an artifact. Paley then compares the watch with biological structures such as the human eye and argues that the eye cannot possibly be the product of any purely natural process. It calls out for a designer, and that designer of course is God.

Even today, Christians are told to look at the beauty and complexity of the world about them, smell the flowers, peer through telescopes into the heavens, and bear witness to God's creative artifacts in nature. The *argument from design* remains the most common *scientific* argument theists give for their beliefs, despite the fact that evolution by natural selection is now solidly confirmed as the primary mechanism by which complex living organisms develop from simpler forms.

EVOLUTION AND GOD

When Charles Darwin entered Cambridge he was assigned the same rooms occupied by William Paley a generation earlier. Darwin was very impressed by Paley's design argument and committed much of it to memory. But ultimately, after years of detailed observations of the natural world, Darwin found the design argument unconvincing. In 1859 Darwin published *On the Origin of Species*, laying out the detailed evidence from a wide range of biological examples that living organisms evolve by a process of random mutations and natural selection. Alfred Russel Wallace (as well as at least two others) had independently discovered this process. He and Darwin announced their conclusions simultaneously. But while Wallace and the others had inferred the correct mechanism, Darwin's case was much stronger based on his vast accumulation of data and his impeccable analysis of those data. Darwin earned his reputation as one of the two or three greatest scientists who ever lived.

Natural selection, as proposed by Darwin and Wallace, was then, and now remains difficult for many people to grasp. It is really quite amazing that an unguided, purely material process can produce the fantastic complexity of living organisms. Yet it did. The evidence is overwhelming in the fossil record and in the DNA that is shared by every cell of all earthly life, from bacteria to redwoods, giraffes, humans, and the strange forms living in deep-sea thermal vents.

Evolution takes a long time and requires that the Earth be much older than six thousand or so years inferred from the Bible. Just as Darwin was beginning his investigations on HMS *Beagle*, Charles Lyell and other geologists had begun to build a case for a much older Earth than implied by the Bible. Indeed, Darwin brought along the first volume of the first edition of Lyell's classic work *Principles of Geology*, ironically a gift from the ship's captain, Robert FitzRoy, a religious zealot who later would deeply regret the role he played in the development of Darwin's theory.

Darwin was deeply disturbed, however, by the calculations of the great physicist William Thomson, Lord Kelvin, which indicated that the lifetime of the sun would be far too short for evolution to proceed on Earth. Kelvin knew of only two possible sources of energy—gravity and chemical reactions—and neither were sufficient. However, no one at the time even knew about atomic nuclei, much less nuclear energy, which were not discovered until the early twentieth century. We can now confirm that the sun is efficiently powered by nuclear fusion reactions at its center that will burn for ten billion years. In the late 1990s I collaborated on an experiment in Japan called Super-Kamiokande that observed the neutrinos from the center of the sun emitted in these reactions.[7]

Evolution completely contradicts the biblical story of the origin of life and *On the Origin of Species* triggered a protracted war between science and those who insist the Bible is the literal word of God. That war has continued unabated to the current day. Polls indicate that only 10–15 percent of Americans can imagine an evolutionary process in which God does not play a role. This may seem to contradict the fact that 20–25 percent of Americans are Catholics and the Catholic Church officially accepts evolution by natural selection. Two popes have affirmed it (the current pope, Benedict XVI, seems to be equivocating).[8] However, the Church has made it clear that God still plays a role as the author of all natural law and does not accept nondirected evolution as an explanation for mind. The naturalistic view of evolution holds that no supernatural intervention is needed for either body or mind and that mind is a product of purely material processes.

A rational dialogue between science and Bible-literal religion is probably impossible, since each proceed from a different presupposition. Scientists presuppose that the scientific method of objective observation and logical or mathematical model building is the best, and perhaps only way to arrive at useful information about the world. Fundamentalists presuppose that their scriptures are divinely inspired and so must not be questioned. If they contradict science, then science is wrong and must be proven so at any cost.[9]

Scientists like myself base our presupposition on the immense technological success of science that even fundamentalists do not hesitate to make use of. There is something absurd about biblical and Qur'anic literalists flying around the world on jets instead of magic carpets, communicating by cell phones and the Internet rather than smoke signals, and publishing antiscience tracts on laser printers rather than carved rocks.

Since we still do not understand how life originally came about on Earth, anyone is free to propose that God initiated the process. However, saying that "God did it" is no more an explanation than "Nature did it." We need to know how it was done. Furthermore, just because science cannot currently explain the origin of life, we have no reason to expect it never will.[10] Seeking God in the gaps of scientific knowledge has never proven to be a fruitful enterprise. Science has always had a way of filling its own gaps.

The mutations that allow for evolutionary change appear to be random with the implication that every species that ever lived, including *Homo sapiens*, is an accident. The famed paleontologist Stephen Jay Gould claimed that if you were to start evolution up all over again without changing anything else, humanity would not have evolved, nor any other species that today walk on the earth, swim in its waters, or fly in its skies. Gould also insisted that there was no guarantee evolution would always act to produce increasingly complex forms of life with ever-broadening capabilities.[11]

However, this remains controversial. We will return to this issue in chapter 15 when we discuss Simon Conway Morris's proposal that evolution in fact converges toward the development of intelligent life and perhaps humanity itself without divine intervention. This implies that a deist god could have made the universe with sufficient uncertainties to allow for both human free will and the natural evolution of humans.

DARWINISM AND DEISM

The Darwinian mechanism for evolution is a combination of chance and natural selection. But at the time of Darwin, science had no place for chance. In the clockwork universe everything is predetermined, and this includes the mutations that trigger evolutionary changes. Chance, in other words, is an illusion in the clockwork universe.

However, Darwinism is perfectly compatible with the deist god of the Enlightenment. Evolution can be seen as the way the deity decided to create life in the universe, the whole development of life including humanity being written into the laws of nature at the creation. The problem is, this requires an acceptance of the clockwork universe and leaves no room for human free will. Based on the science alone, as it was known prior to the twentieth century, free will is impossible and the deist view remains the best choice.

However, as we will see in later chapters, the clockwork universe has been refuted by quantum physics, thereby pulling the rug out from Enlightenment deism. This allows for the possibility of free will, but also requires that God step in countless times during the process of evolution to guarantee that the human species will appear. We will see how contemporary theologians are attempting to find a way that God can still act in the universe consistent with known science.

NOTES

1. Paul Leicester Ford, ed., *The Writings of Thomas Jefferson* (New York: Putnam, 1892–99), 4: 430.

2. The early part of the Age of the Enlightenment is also called the *Age of Reason.*

3. Karen Armstrong, *The Battle for God* (New York: Ballantine Books, 2000), p. 77.

4. For the history of deism I have relied on and paraphrased heavily from Will and Ariel Durant, *The Story of Civilization: Part IV, The Age of Voltaire* (New York: Simon & Schuster, 1965), pp. 119–21. I have also referred to the article on deism by Bill Cooke in *The New Encyclopedia of Unbelief*, ed. Tom Flynn (Amherst, NY: Prometheus Books, 2007).

5. For more details, see Armstrong, *The Battle for God.*

6. William Paley, *Natural Theology or Evidences of the Existence and Attributes of the Deity Collected from the Appearance of Nature* (London: Halliwell, 1802).

7. For a picture of neutrinos emitted from the sun, see http://apod.nasa.gov/apod/ap980605.html (accessed June 30, 2008).

8. Pope John Paul II, "Address to the Academy of Sciences, October 28, 1986," *L'Osservatore Romano*, English ed., November 24, 1986, p. 22.

9. Barbra Forrest and Paul R. Gross, *Creationism's Trojan Horse: The Wedge of Intelligent Design* (Oxford: Oxford University Press, 2004).

10. Richard Robinson, "Jump Starting a Cellular World: Investigating the Origin of Life, from Soup to Networks," *Public Library of Science Biology* 3, no. 11 (November 2004): e396, http://biology.plosjournals.org/perlserv/?request=get-document&doi=10.1371%2Fjournal.pbio.0030396 (accessed June 30, 2008).

11. Stephen Jay Gould, *Wonderful Life: The Burgess Shale and the Nature of History* (New York: Norton, 1989).

8

THE SPOOKY QUANTUM

We are not only observers. We are participators. In some strange sense, this is a participatory universe.

—John Archibald Wheeler[1]

LIGHT IS A WAVE

Let us take a first look at what it is about modern physics that gives the new spiritualists so much confidence that physics has opened up for them a world beyond matter. We begin with the strange behavior of light.

One of the most familiar yet mysterious physical phenomena is light. At first glance, light would seem to be some kind of "pure energy" with qualities that almost border on the supernatural. In Genesis, the first thing God says when he creates the universe is "Let there be light." Our language abounds with expressions that use light as a metaphor for ultimate knowledge, both worldly and divine. When explaining something we "throw light" on the subject, a good lecture is "illuminating," and as we gain wisdom we attain "enlightenment."

In the seventeenth century, Isaac Newton proposed that light was

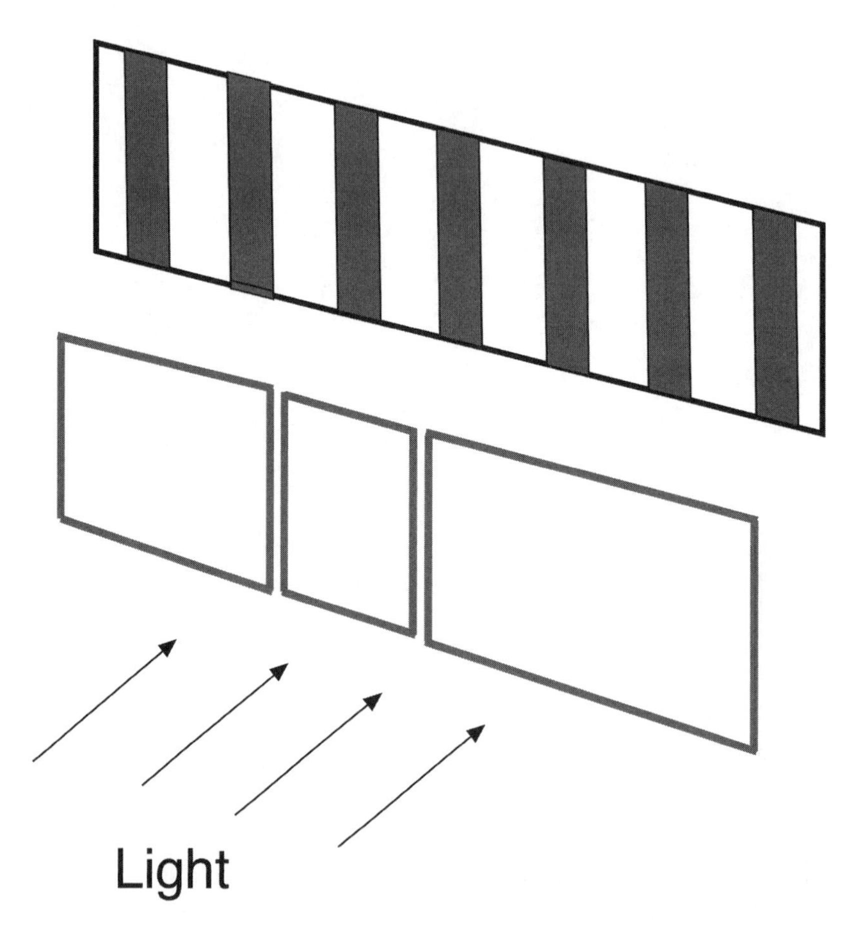

Fig. 8.1. Young's double-slit experiment. Monochromatic light from the left shines on two slits in an opaque surface. An alternating pattern of dark and light bands is seen. The light bands are actually brightest in the center, with their intensities falling off toward the edges.

composed of particles. He rejected the alternate proposal of Christiaan Huygens (d. 1695) that light was a wave phenomenon, comparable to the sound vibrations observed in materials such as water and air, because light did not bend around corners the way waves can. We can generally hear around corners, while it seems we can't see what hides behind. Newton did not look close enough. Light indeed does bend around corners, ever so slightly, as can be seen by holding a card pierced with a tiny pinhole up to a lamp.

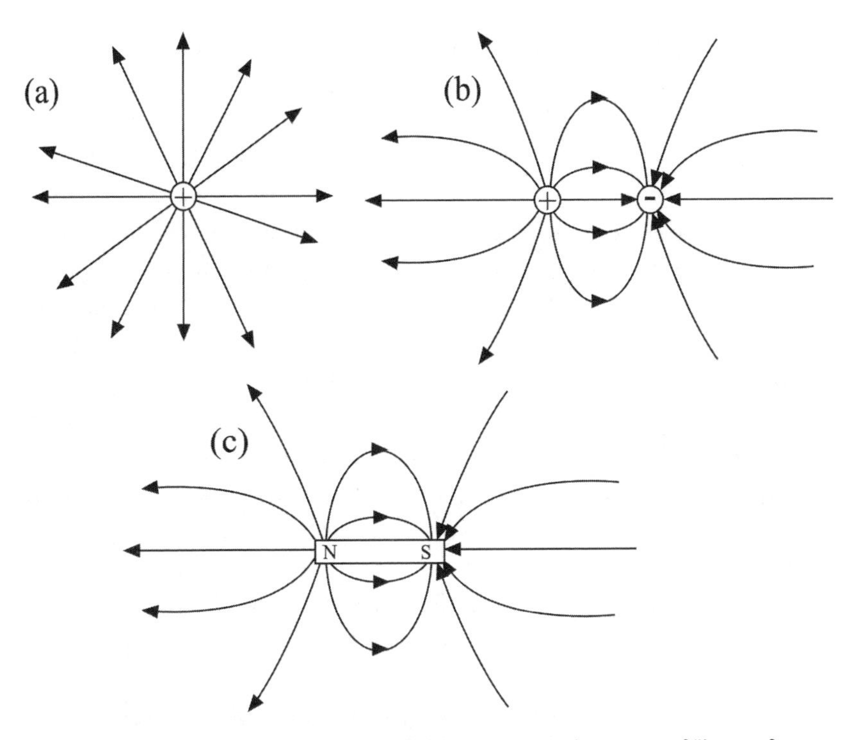

Fig. 8.2. Electric and magnetic fields are pictured in terms of "lines of force," In (a), the electric field surrounding a positive point charge is shown. In (b) the electric field surrounding an *electric dipole* composed of two equal and opposite point changes is presented. In (c) we see that the magnetic field surrounding a bar magnet is similar to (b), which is called a *magnetic dipole*. These are all just visualizations. No one has ever "seen" an electric or magnetic field.

In 1800 Thomas Young (d. 1829) passed a beam of light through two narrow slits in an otherwise opaque screen and observed alternating bands of light on a white screen, as shown in figure 8.1. Using a ripple tank to investigate how water waves emerge from two openings in a barrier, he was able to explain the effect as the pattern of interference between light waves emerging from each slit. Since then science has treated light as a wave phenomenon.

Until the twentieth century, gravity, electricity, and magnetism were the only known forces in nature. We often talk about "contact" forces, such as when you push a box across a surface, or frictional forces between the box and the surface that resists that push. However, when you examine these at the atomic level you see they are simply electrical repulsions that result as you try to move the clouds of electrons in the atoms in, say, your hand, against the cloud of electrons in the atoms in the surface of the box.

The three basic forces are often visualized in terms of invisible "fields" that emanate from some source and reach across space to effect an interaction with another body, as illustrated in figure 8.2. The source of the electric field is a region of electrical charge. The source of the magnetic field is electric current—a flow of electric charge. The bar magnet shown in figure 8.2 (c) is equivalent to a coil of wire with an electric current flowing through it. Inside the iron of the bar magnet, electrical currents frozen into place produce the net magnetic field that is measured.

The electric field lines, or "lines of force" in the first two figures are determined by placing a positive pointlike test charge at various places in space and following the path of its motion, the arrow giving the direction of acceleration. The magnetic field lines can be similarly mapped out by placing a small compass needle at various places and tracing the directions it points.

Note that in neither case is the field line directly observed. Rather, what is drawn is a model inferred from measurements with observable bodies. We have no evidence that these fields exist in "objective reality." They are simply pictures physicists draw to help them predict the motion of charged particles and currents placed in the vicinity of other charged particles and currents.

We have seen how in the mid-nineteenth century Michael Faraday showed that the electric and magnetic forces represented a single phenomenon called *electromagnetism.* He demonstrated that a time-varying electric field produces a magnetic field and vice versa. This *principle of electromagnetic induction* provides the basis for electric motors and generators.

We have seen that the unification of the forces of nature began when Newton showed that the forces of gravity in the heavens and that on Earth are the same. This program continues today with the ultimate goal being able to show how all the forces of nature arise from one single universal force.

In 1865 James Clerk Maxwell (d. 1879) wrote down four equations, now called *Maxwell's equations*, which described both the electric and the magnetic fields in a unified way. Maxwell showed that the equations allowed for the existence of *electromagnetic waves* in empty space. These waves were predicted by the theory to move at exactly the measured speed of light, $c = 300,000$ kilometers per second in a vacuum. An obvious inference was that light itself is some kind of electromagnetic wave. This conclusion was confirmed in 1887 when Heinrich Hertz (d. 1894) generated radio waves in the laboratory and showed that they moved at the speed of light.

In the case of sound, the waves that are propagated consist of pressure and density vibrations in a medium such as air or water. In the case of electromagnetism, what is doing the waving? It was conjectured that a transparent, frictionless medium called the *ether* pervaded all of space and electromagnetic waves propagated as the vibrations of the ether.

However, this presented a problem. The observed speed of water waves depends on the speed of the receiver relative to the waves. Paddling a surfboard out from shore at Waikiki Beach in Hawaii, you will experience the waves moving toward you at a greater speed than when you paddle toward shore.

It stood to reason, then, that the speed of light should change as Earth moved through the ether. In 1897 the American physicists Albert Michelson and Edward Morley tried to measure changes in the speed of light as Earth turned in its orbit around the sun, changing its direction through the ether. To their surprise, they found no differences. The speed of light did not depend on motion through the ether.

Oddly enough, this was actually consistent with Maxwell's equations, which said that the speed of light in a vacuum was always c and made no provision for motion through the ether.

EINSTEIN'S RELATIVITY

In 1905 a young physics PhD working as a patent clerk in Bern, Switzerland, named Albert Einstein asked what the consequences would be if the speed of light did not depend on the motion of the light source or the observer. The result, presented in a theory called *special relativity*, revolutionized our concepts of space, time, matter, and energy.

Einstein showed that our commonsense notions of space and time, in which they form some kind of absolute framework in the universe with respect to which material objects move, are grossly wrong. Rather, our

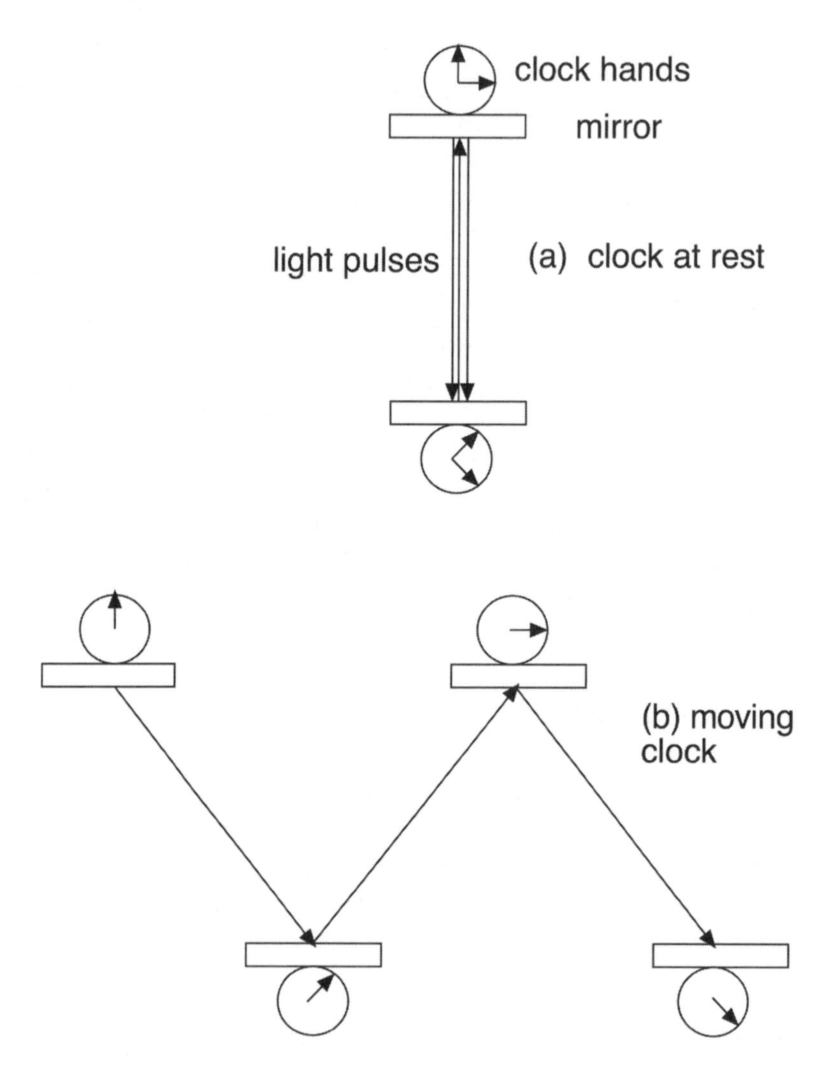

Fig. 8.3. A light pulse clock. Each time a pulse hits a mirror, the hand on the clock attached to that mirror moves one tick. The top clock starts at zero and one tick is shown. Two ticks on the bottom clock are shown. The hands show the position at each tick. In (a) the clock is at rest in the observed reference frame. The pulse paths are displaced for clarity. In (b) the clock is moving in the observer's reference frame. The distance traveled by the light pulse between mirrors is longer, and since the speed of light is the same in all reference frames, the moving clock runs slower.

measurements of spatial and temporal intervals are relative—they depend on the observer's point of view, or what we call a *frame of reference.* In an effect called *time dilation*, a moving clock will appear to run slower than a clock at rest. That's not to say that the clock is actually running slower. It runs normally in its own frame of reference, the one where it is at rest, but it runs slower to an observer in another reference frame in which the clock is moving. In a related effect called the *Fitzgerald-Lorentz contraction*, a moving object will appear shortened in the direction of its motion.

To see how this comes about, imagine a clock composed of two mirrors (see figure 8.3). A light pulse moves back and forth between the mirrors, causing a "tick" every time a mirror is hit. When the clock is moving with respect to the observer, the light must travel a longer, diagonal distance. But since light always moves at the same speed (a vacuum is being assumed here), it takes longer to go between mirrors in the moving reference frame and so the tick rate slows down for the observer watching the clock move by. This is time dilation.

The implications are profound. For example, suppose we could build a spacecraft that was able to leave Earth at a constant acceleration equal to one *g*, the acceleration of gravity on Earth (providing artificial gravity to the crew), it could travel to the neighboring galaxy, Andromeda, which is 2.4 million light-years away (accelerating the first half of the trip and then decelerating the second half), in only 30 years' elapsed time as measured on the clocks aboard the ship. If the ship turns around, the astronauts will return home aged by 60 years, while almost 5 million years will have passed on Earth.

While such an adventure, of course, has not been made, many other tests, both with very high-speed particles and with highly precise atomic clocks at jetliner speeds, have confirmed the validity of time dilation.

Fitzgerald-Lorentz contraction follows from time dilation. Suppose the distance from Earth to a star is 100 light-years and astronauts are traveling to the star at a constant 0.99 times the speed of light and it takes them 14 years to get there (rather than the 101 it would without time dilation). The distance the astronauts measure is their velocity, 0.99 light-years per year, times the time of the trip to get there, 14 years, or 13.9 light-years. It is as if in their reference frame they watched a meter stick 100 centimeters long going by at 0.99 c (c being the speed of light in a vacuum). In that reference frame the stick would appear 13.9 centimeters long.

One of the important results of the special theory of relativity is that no object can move faster than the speed *c*. The Stanford Linear Accelerator in Palo Alto, California, is able to accelerate electrons at 0.9999999997 of the speed of light but not beyond. Photons and other massless particles travel at exactly the speed of light in a vacuum.[2]

In his *general theory of relativity*, published in 1916, Einstein also showed that light is affected by gravity. This has been confirmed by observations during complete solar eclipses of the bending of light from a star near the edge of the sun.

The conclusion is that light is *not* some kind of substance separate from matter. Light is matter, with all its inertial and gravitational properties.

LIGHT IS A PARTICLE

Despite its astounding success, the nineteenth-century wave theory of light could not explain several other observations. Let us start with what is called *blackbody radiation.*

Every physical object emits electromagnetic radiation with a smooth wavelength spectrum that depends on the temperature of the body. In the case of objects from everyday experience, such as a rock or a human body, that radiation is in the infrared region of the spectrum where the wavelengths are larger than those of visible light and not detectable by the human eye. So in the absence of any reflected light, it appears to the naked eye that the body is black. For example, the spectrum of radiation from a human body at 37 degrees Celsius (98.6 degrees Fahrenheit) is shown in figure 8.4. The term "blackbody" radiation is applied to the phenomenon, although sufficiently hot objects such as the sun or a burner on an electric stove will radiate in the visible region.

The wave theory of light implied that the radiation would be increasingly intense as one went to shorter and shorter wavelengths. This is easy to see if you imagine a metal box filled with waves. The shorter the wavelength, the more waves can fit in the box. This theoretical prediction of the wave theory of light disagrees with the observed spectrum, which indicates a smooth drop to zero in the wave intensity as wavelength decreases.

In 1900 Max Planck was able to accurately describe the blackbody spectrum by assuming that light occurs in discrete bundles of energy. The

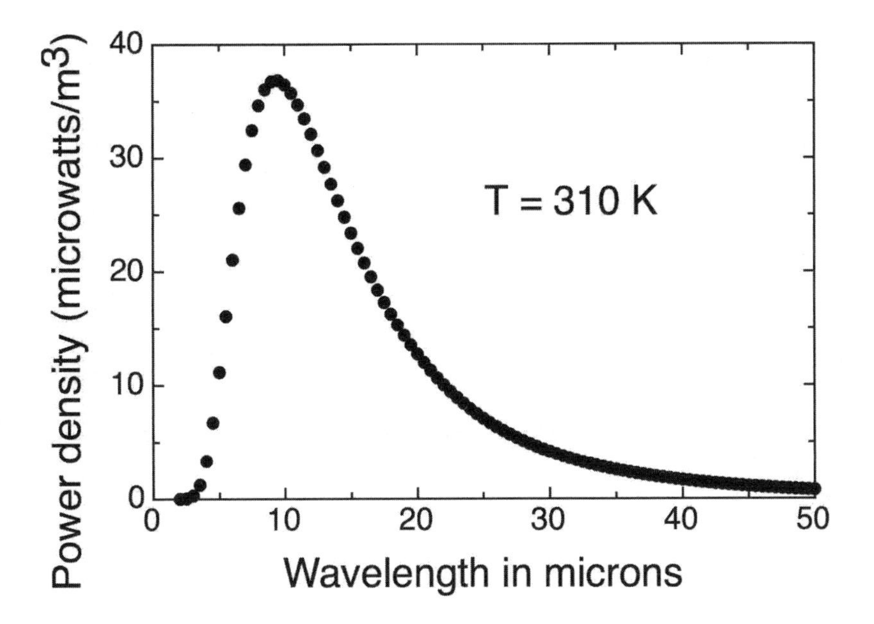

Fig. 8.4. The Planck spectrum for a blackbody at a temperature of 310 Kelvin (37°C or 98°F, the temperature of a human body). The wave theory predicts an ever-increasing intensity at shorter wavelengths, which is not observed.

energy of each bundle was composed of an integral number of basic units of energy called *quanta*, where the energy of each quantum was proportional to the frequency, or inversely proportional to the wavelength, of the electromagnetic wave associated with the light. The constant of proportionality, designated by *h*, we now call *Planck's constant.* Energy conservation smoothly cuts off the spectrum at the shorter wavelengths that correspond to higher energies. Thus was born quantum physics.

A second phenomenon that classical wave theory could not explain was the *photoelectric effect*, in which ultraviolet light, that is, light with wavelengths shorter than visible light, produced an electric current when directed on certain metals. At the time this effect was not seen for visible light no matter how intense it might be, while even very weak ultraviolet light produced a current.

In 1905, the same year he introduced special relativity, Einstein explained the photoelectric effect (see figure 8.5). He postulated that the fundamental quanta of light discovered by Planck were actual physical particles. These were later dubbed *photons*. When a photon of sufficient energy struck metal, it would kick out an electron, which then produced an electric current. Visible photons generally do not have sufficient energy to lift an electron out of the metal, while the photons associated with ultraviolet light have higher energy. Einstein was able to quantitatively describe the effect using Planck's relationship between energy and frequency with the same constant, *h*.

Incidentally, the interpretation of light as a beam of particles did away with the need for an ether to do the vibrating of an electromagnetic field that constituted a light wave. But this did not change the fact that light still behaves like a wave, bending around corners and producing interference patterns in a double-slit experiment. Light is a particle. But it still looks like a wave.

PARTICLES ARE WAVES

The momentum of a photon is inversely proportional to the wavelength of the corresponding electromagnetic wave, with Planck's constant *h* the constant of proportionality. This is verified in experiments such as the *Compton*

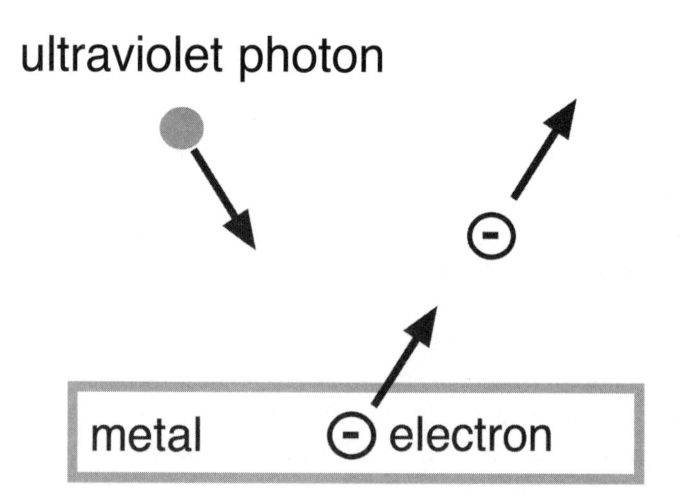

Fig. 8.5. The photoelectric effect. Ultraviolet photons kick electrons out of metals, where they are relatively loosely bound, which then produce an electric current.

effect, in which x-ray photons are scattered off electrons and their loss of momentum is seen as a decrease in wavelength. In 1922, Louis de Broglie proposed that this same relationship holds for all particles, such as electrons, and that particles should exhibit similar wave effects. This was verified five years later in a laboratory study of electron interference.

So, all material bodies, whether photons, electrons, protons, rocks, planets, or you and I, have both particlelike and wavelike properties. Whether or not an object behaves like a wave depends on its wavelength. We can hear sound around a corner because the wavelengths of sound waves are comparable to the size of everyday objects, the audible range going from about 0.2 to 20 meters. On the other hand, the wavelengths of visible light range from about 0.4 to 0.7 micrometers, where a micrometer is a millionth of a meter. So we generally can't see around corners, although light diffraction can be observed with the naked eye if the light passes through a tiny pinhole.

An electron moving at 500 meters per second has a wavelength of 1.5 microns. A beam of such electrons passing through a hole with a diameter of 10 microns will spread out by about 8 degrees. By comparison, a 50-kilogram woman running at 5 meters per second has a wavelength of 3×10^{-36} meter. That's why you can't see her coming around a corner. In short, wave effects for familiar bodies are too small to be noticed but can be quite pronounced at the atomic and the subatomic levels.

THE WAVE FUNCTION

In 1925 Werner Heisenberg developed the first formal version of mathematical quantum theory called *matrix mechanics*. An alternate version called *wave mechanics* was developed a year later by Erwin Schrödinger. In the classic book *Principles of Quantum Mechanics*, first published in 1930 and still the definitive textbook on the subject (I have the 1989 printing on my shelf), Paul Dirac unified the two theories in an elegant formalism he called *transformation theory* that also enabled him to make the theory relativistic, that is, valid for particles moving at or near the speed of light.

Schrödinger wave mechanics is the most familiar since it is the one using the mathematics of partial differential equations that is taught at the undergraduate level. Matrix and linear vector algebra are a bit more advanced and general in their applications.

In wave mechanics the state of a particle is defined by a complex mathematical function called the *wave function* that depends on spatial position and time. Max Born proposed that the square of the magnitude of the wave function gave the probability for finding the particle in a tiny unit volume at that position and time. The Heisenberg and Dirac formalisms both reduce to the same conclusion as Schrödinger mechanics, namely, that quantum mechanics does not predict the motion of individual bodies but only the behavior of a statistical ensemble of similarly prepared bodies.

THE UNCERTAINTY PRINCIPLE

Perhaps the most important innovation in quantum mechanics was the uncertainty principle, introduced in 1927 by Heisenberg, which says that the momentum and the position of a body cannot be simultaneously measured with unlimited precision. (Except for speeds near the speed of light, momentum can usually be approximated as the product of the mass and the velocity of a body.) The uncertainty principle can be proved mathematically in any of the three formalisms described in the preceding section.

The Heisenberg Uncertainty Principle

The product of the uncertainty[3] in position and uncertainty in momentum of a particle is greater than or equal to Planck's constant divided by 4π.

We can understand the uncertainty principle from two established facts. First, in order to measure the position of a body with a given precision you have to bounce a photon or other particle with wavelength less than the desired precision off that body. Second, the lower the wavelength, the higher the momentum of the particle will be. Some or all of that momentum will be transferred to the object being studied, making its momentum uncertain by that amount. Thus, the more precise you try to measure the position, the less precise you know the body's momentum. You can never know each to unlimited precision.

Newton had provided us with laws of motion (see chapter 6) that enabled physicists to predict the motion of bodies with what seemed, in principle, to be unlimited precision. This implied that the universe itself is one vast machine, a clockwork universe in which everything that happens is completely predetermined by what went on before. However, such a prediction requires a knowledge of both the position and the momentum of the body with unlimited precision. Heisenberg refuted this notion, showing that an inherent uncertainty exists in the motion of bodies and the best we can do is predict their average motion.

For example, there is no way for a physicist to predict with any reasonable accuracy the motion of an unbound electron initially located within a volume the size of an atom but to that atom. The uncertainty in the electron's velocity by virtue of its position being so well known is one million meters per second with random direction! Six seconds later the electron can be anywhere within a volume the size of Earth. By contrast, the uncertainty in the velocity of a body of mass equal to one gram confined to a cubic centimeter is 5×10^{-30} meter per second and the motion of such a particle can be predicted with great accuracy.

The same equations of motion that appear in classical Newtonian mechanics can be used to predict the average motion of an ensemble of particles, but not that of individual particles. This is not to say that the motions of particles are completely random. Quantum mechanics has what is called *statistical determinism.* The motions of particles are constrained to yield the calculated average the same way the toss of a coin is restricted to give an average of half heads and half tails. Indeed, the exact statistical distribution giving the range of deviations from average motion can also be calculated.

The uncertainty principle will play an important role in our later discussions of "quantum theology." For now, let us just note that it affirms the statistical nature of quantum mechanics.

WAVE-PARTICLE DUALITY

If relativity violated common sense, quantum mechanics did so even more. Although it developed into a theory of incredible precision, one that to this day agrees (as does relativity) with every observation ever made, people still argue about what it "really means." This has left quantum mechanics open to interpretations supporting unconventional claims that are not accepted by mainstream science.

Let us consider a particularly puzzling experiment, which I first heard Richard Feynman talk about in 1956 in a special course for engineers at Hughes Aircraft Company in Culver City, California, where I was working at the time. He was describing Young's double-slit experiment using electrons instead of photons, but the consequences would have been the same either way. He added an electron detector behind one slit (see figure 8.6). Let us assume for the purposes of this discussion that we can detect the electron without seriously deflecting it from its path or taking away significant amounts of its energy.

If the detector is on, then we know which slit the electron passed through. After illuminating the slits with many electrons using a beam of light, we get two bright bands on the wall. One band will be for the light

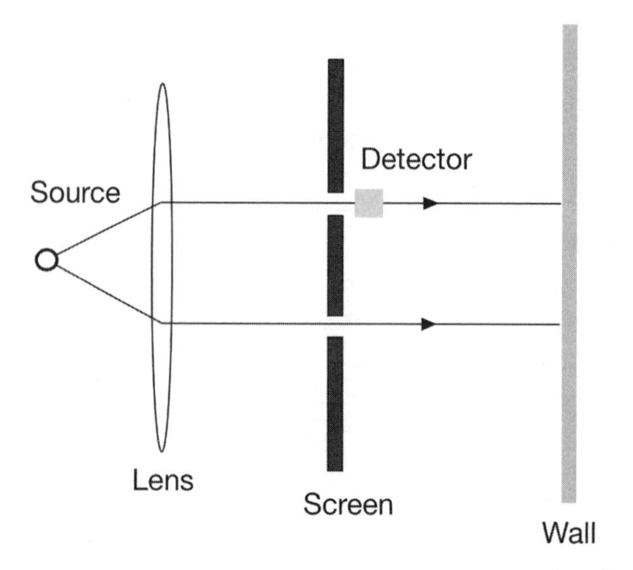

Fig. 8.6. In the double-slit experiment, a detector (D) is placed behind one slit and turned on or off at the discretion of the experimenter. When D is off the usual interference pattern of bright and dark bands is seen, as in figure 8.1. When D is on the pattern goes away and two bright bands are seen as expected for separate beams of particles from the two slits. The purpose of the lens is to produce parallel beams.

passing through the slit as registered by the detector. The second band will be for the electrons that do not register in the detector and so must have passed through the other slit. In short, we get no interference effect.

When the detector is off the interference pattern appears. We do not know whether the electron went though one slit or the other, which would be the case if we viewed it as a wave rather than a particle. In the wave picture, the wavelets emanating from each slit came from the same initial wave and these interfere with one another to produce the observed pattern.

So is the electron a particle or is it a wave? At first glance, whether an object is a wave or a particle seems to depend on what you decide to measure. If you measure a wave property such as interference, then the object is presumably a wave. If you decide to measure a particle property such as position, then the object is presumably a particle. This is called *wave-particle duality.*

To make this even more spooky, we can set up the experiment so the decision to measure a wave property or particle property is made after the object leaves the source. That source can be light from a galaxy 10 billion light-years away. So, it would seem, we not only can control reality with our conscious minds, we can do so over a distance equal to the size of the visible universe and 10 billion years or more back in time.

As we will see, the wave-particle duality is the primary basis for the claims of the new spiritualists that we can make our own reality.

SPOOKY ACTION AT A DISTANCE

Let us look at another example. Figure 8.7 illustrates the process of detecting an electron and how it is described in terms of the quantum wave function. Before detection the position of the electron passing through the screen is known to the accuracy of the width of the slit. It can be anywhere in that slit, so its wave function—which gives the probability for finding the electron in a particular region of space—is spread out in space about that amount. The electron is then observed in detector A, thus locating it in space more accurately. At the moment of detection the wave function instantaneously "collapses" to the size of A.

Suppose the screen is absent and we start out knowing nothing about the position of a particle. Then the particle's wave function is in some sense spread throughout the universe. It has the same magnitude at every spatial point. Then when a measurement is made, the particle's position becomes known to be in some small region the size of the detector and the wave function collapses to that size. Einstein called this a "spooky action at a distance," since the collapse happens instantaneously throughout the universe. Again, it would seem that the act of conscious measurement has reached out in space at infinite speed to the farthest corner of the universe.

THE EPR EXPERIMENT

In 1927 Louis de Broglie, who, five years earlier had successfully predicted that particles such as electrons would exhibit wavelike behavior, proposed the idea that the wave function was a kind of *pilot wave* that deterministically guides a particle along its path. While Einstein, who never accepted the statistical nature of quantum mechanics ("God does not play dice")

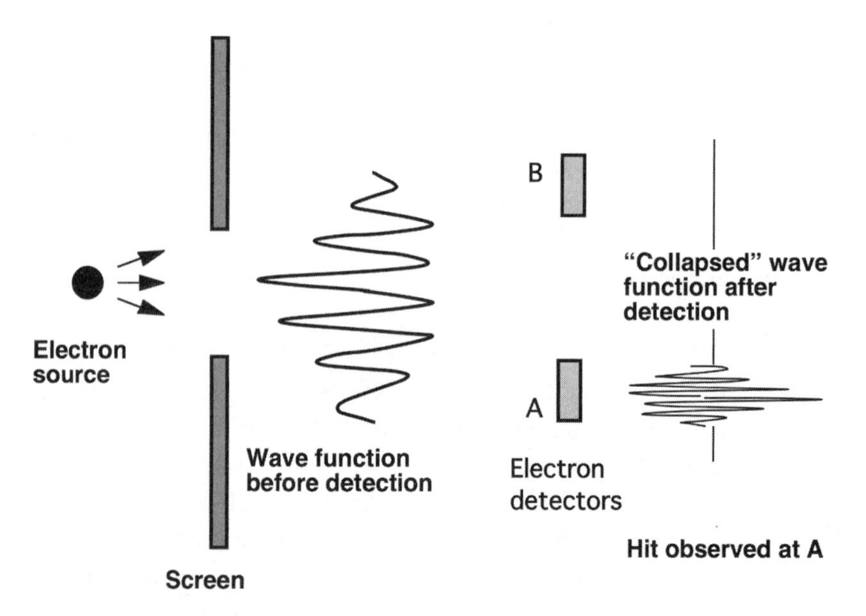

Fig. 8.7. An illustration of what Einstein called "spooky action at a distance." An electron passes through an aperture. Its wave function is then approximately the same size as the aperture. After detection at A the wave function "collapses" to the size of detector A and is zero everyplace else in the universe.

liked the idea, de Broglie received little other support, in fact some derision, and he did not pursue it further.[4]

In 1935, Einstein and two younger colleagues, Boris Podolsky and Nathan Rosen, wrote a paper showing by means of a thought experiment (*gedankenexperiment*) that quantum mechanics is either "nonlocal," that is, allows for influences that move through space faster than the speed of light, or is not a complete theory, that is, there are missing elements yet to be discovered.[5] If quantum mechanics is to remain a "local" theory, obeying Einstein's speed limit of the speed of light, then it is just an approximate theory awaiting discovery of the final subquantum theory that lies behind it. This became known as the *EPR paradox.*

Despite Einstein's great prestige, subquantum theories were thought impossible until 1952 when David Bohm rediscovered a simple mathematical fact that had been published in 1926 by Erwin Madelung.[6] He showed, as had Madelung, that the quantum mechanical Schrödinger equation, which is used to compute the wave functions and energy levels for nonrelativistic systems such as atoms and harmonic oscillators, can be rewritten as a classical equation of motion, the Hamilton-Jacobi equation, with the addition of a term Bohm called the *quantum potential.*[7] The quantum potential depends on the shape but not the magnitude of the wave function. This suggested to him that the underlying principles of particle motion were deterministic, just as in Newtonian mechanics, and that motion was guided by "hidden variables" that were not directly observed.

As it developed, the notion of hidden variables was more than just another purely philosophical interpretation of quantum mechanics with no unique empirical consequences. In 1964 theoretical physicist John Bell showed that if hidden variables, whatever their form, are local, then they could be tested against conventional quantum mechanics.[8] The test involved an experiment that Bohm had proposed in his excellent 1950 textbook, *Quantum Theory*, as a practical way to implement the thought experiment proposed by EPR.[9]

Bohm's proposed experiment involved electrons. However, I will describe the equivalent experiment done with photons, which, as we will see below, was how the test was eventually carried out definitively.

As illustrated in figure 8.8, pairs of photons are emitted in opposite directions from a source. The total spin (intrinsic angular momentum) of each pair of photons emitted is zero. This is called a *singlet state.* Since a

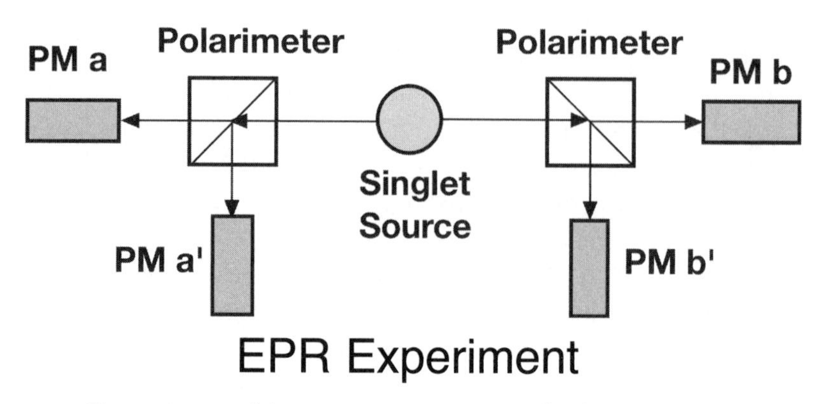

Fig. 8.8. Layout of the EPR experiment. See text for details.

single photon has one unit of spin, the two photon spins are equal and opposite.

Each photon travels to a linear polarimeter that can be adjusted to send the photon toward either of two photomultiplier (PM) tubes that can detect individual photons. The polarimeters are like fences with parallel spaces that let through electromagnetic waves with a electric field component in that direction. (The wave function of a photon is an electromagnetic wave.) The axes of the polarimeters are shown in figure 8.9, which is in the plane perpendicular to the photon direction. The angle θ is adjustable.

Bell proved that for any quantum theory of nonlocal hidden variables, a certain quantity *S* that measures the correlation between four photon spin components along the four axes ab, ab', a'b, and a'b' must have a magnitude of less than two.[10] This is shown by the shaded region in figure 8.10. We see that classical mechanics, as given by the dashed curve, also obeys Bell's theorem. As Bell proved, however, conventional quantum mechanics violates his theorem, predicting the solid curve shown for the variation of *S* with angle.

Since *S* is a measure of how strongly the two spins are correlated, quantum mechanics predicts a greater correlation than classical mechanics. Any value greater than 2 in magnitude implies that the photon spins are more correlated, that is, more dependent on each other than they would be if each photon had a definite spin that is independent of measurements at the end of each beam line.

Let me try to explain. Quantum mechanics says that the spin states of

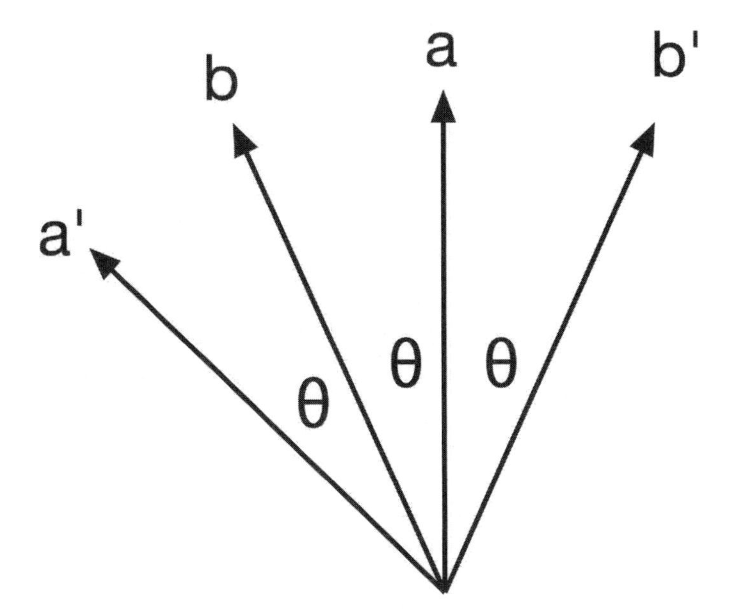

Fig. 8.9. Alignments of polarimeter axes in EPR experiment.

the photons in a spin zero pair are "entangled." That means that they do not have definite spin orientations. When the component of spin of the photon at the end of one beam line is measured along a certain direction, then since the total spin must be zero the component of spin of the other photon along that direction must be equal and opposite. Somehow the second photon "knows" the result of the measurement of the first photon, even though a signal between the two ends of the beam line would have to travel faster than the speed of light.

In 1982 a definitive series of "EPR experiments" with this configuration was carried out by Alain Aspect and his collaborators at the Institut d'Optique Théoretique et Appliquée in Orsay, France.[11] They arranged it so the decision on the orientation of the polarimeter axes was made *after* the photon pair left the source so that only a superluminal signal could share the information between the detectors. The results agreed perfectly with conventional quantum mechanics and thus ruled out any subquantum theory with local hidden variables.

Again, we find that a physical quantity does not have a definite value in conventional quantum mechanics until it is measured. The photon pair

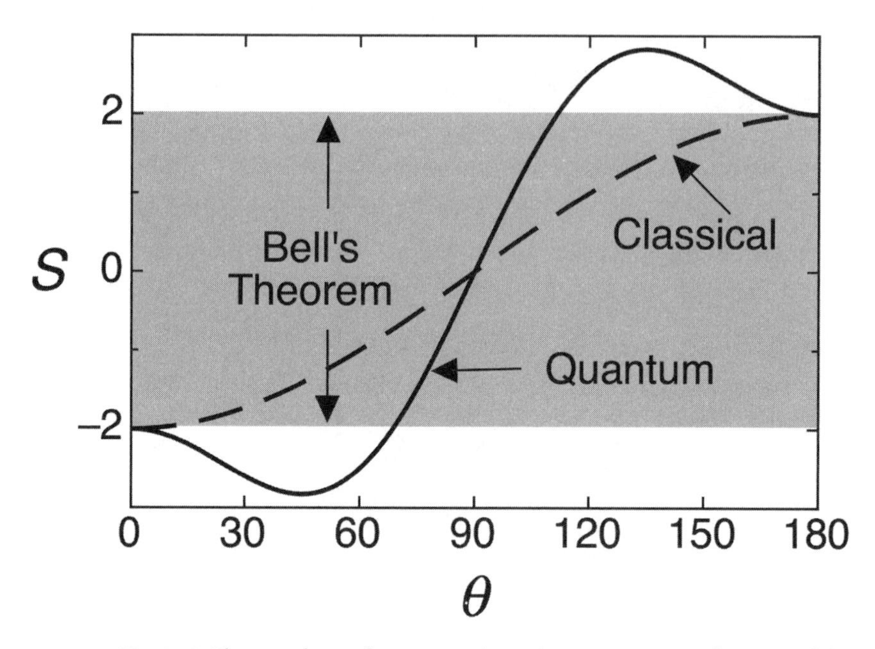

Fig. 8.10. The correlation function in the EPR experiment as a function of the angle between polarizer axes (see figure 8.9). The dashed curve shows what is expected from classical physics. The shaded area is the range allowed by Bell's theorem for a quantum theory with local hidden variables. The solid curve is the prediction of conventional quantum theory, which we see violates Bell's theorem. The data agree precisely with conventional quantum theory, thus ruling out any theory with local hidden variables.

retains its identity as an inseparable whole while the individual photons do not exist in pure spin states until they are measured. It is important to realize that no superluminal signal is implied in this case. Since the two-photon state is one inseparable whole, it has no parts that can signal each other. Indeed, it can be proved in conventional quantum mechanics that it is impossible to use an EPR setup as a superluminal communicator.[12] In the case of hidden variables, the parts do retain their identity as pure states and so a superluminal signal would be required to give the observed result.

The required superluminality of any theory of hidden variables did not discourage Bohm and his supporters. Bohm's model, they reasoned, is simply nonlocal. The quantum potential extends throughout the universe and describes the net contribution of all the other particles in the universe to a given particle's motion. As such it acts instantaneously, faster than the speed of light, violating the principle imposed by Einstein's theory of special relativity that limits all motion and signaling to the speed of light or less.[13] A complete, technical presentation of the Bohm model can be found in Bohm's 1993 book, *The Undivided Universe*, coauthored with Basil J. Hiley.[14]

Bohm offered the new spiritualists yet another reason to believe that the universe is one undivided whole. As we saw in chapter 3, physicist Fritjof Capra claimed that quantum mechanics demonstrates that the universe cannot be broken down into parts but must be treated holistically. Amit Goswami saw heaven in this life, "not a place, but an experience of living in quantum nonlocality."[15]

David Bohm himself became increasingly mystical as he pondered the implications of his nonlocal theory.[16] In a 1979 popular book on the new physics, *The Dancing Wu Li Masters*, Gary Zukav goes into some detail on Bohm's philosophical ideas.[17] Zukav would become a major self-help guru in his own right, with frequent appearances on *Oprah* and a best-selling book, *The Seat of the Soul*. According to Zukav, the soul "has no beginning and no end but flows toward wholeness."[18]

It should be kept in mind that the results of the EPR experiments do not imply that quantum mechanics is necessarily nonlocal, just that any subquantum forces or hidden variables must be so. This is a widely misunderstood fact, even among physicists.

While Bohm's theory continues to have some supporters, most physicists consider the fact that it violates special relativity, rendering it fatally flawed. It will be taken seriously only if superluminal effects are observed.

The EPR experiment results are widely discussed in the literature of quantum spiritualism. Physicists, on the other hand, are underwhelmed. Quantum mechanics has passed yet another empirical test. Ho hum.

NOTES

1. John A. Wheeler in *Mind in Nature*, ed. Richard Q. Elvee (New York: Harper and Row, 1982), p. 1.

2. Technically, particles called *tachyons* that *always* travel faster than light can exist, but none have been observed.

3. Uncertainty is defined as the statistical standard deviation, or the standard error.

4. Victor J. Stenger, *The Unconscious Quantum: Metaphysics in Modern Physics and Cosmology* (Amherst, NY: Prometheus Books, 1995), p. 67.

5. Albert Einstein, Boris Podolsky, and Nathan Rosen, "Can the Quantum Mechanical Description of Physical Reality Be Considered Complete?" *Physical Review* 47 (1935): 777–80.

6. Erwin Madelung, "Quantentheorie in Hydrodynamischer Form," *Zeitschrift für Physik* 43 (1927): 354–57.

7. David Bohm, "A Suggested Interpretation of Quantum Theory in Terms of 'Hidden Variables,' I and II," *Physical Review* 85 (1952): 166.

8. John S. Bell, "On the Einstein Podolsky Rosen Paradox," *Physics* 1, no. 3 (1964): 195–200. For a complete set of Bell's papers on this subject, see John S. Bell, *Speakable and Unspeakable in Quantum Mechanics* (Cambridge: Cambridge University Press, 1987).

9. David Bohm, *Quantum Theory* (New York: Prentice Hall, 1951), pp. 614–23.

10. The mathematical details are worked out in simple form in Stenger, *The Unconscious Quantum*, pp. 112–16 and will not be repeated here. Full references can be found in that volume.

11. Alain Aspect, Phillipe Grangier, and Gérard Roger, "Experimental Realization of the Einstein-Podolsky-Rosen *Gedankenexperiment*: A New Violation of Bell's Inequalities," *Physical Review Letters* 49 (1982): 91–94; Alain Aspect, Jean Dalibard, and Gérard Roger, "Experimental Tests of Bell's Inequalities Using Time-Varying Analyzers," *Physical Review Letters* 49 (1982): 1804–1809.

12. In Stenger, *Unconscious Quantum*, p. 138, I show the results of a computer simulation in which I attempted to transmit superluminal signals with an EPR apparatus. It does not work.

13. John S. Bell, "Introduction to the Hidden-Variables Question," in *Speakable and Unspeakable in Quantum Mechanics*, ed. J. S. Bell (Cambridge: Cambridge University Press, 1987), pp. 29–39.

14. David Bohm and Basil J. Hiley, *The Undivided Universe: An Ontological Interpretations of Quantum Mechanics* (London, New York: Routledge, 1993).

15. Amit Goswami, *The Self-Aware Universe* (New York: Penguin, 1993), p. 263.

16. David Bohm, *Wholeness and the Implicate Order* (London: Routledge and Kegan Paul, 1980).

17. Gary Zukav, *The Dancing Wu Li Masters: An Overview of the New Physics* (New York: Morrow, 1979), pp. 323–31.

18. Gary Zukav, *The Seat of the Soul* (New York: Fireside, 1989), p. 37.

9

THE ELEMENTS OF MATTER

If you were to measure the distance from Los Angeles to New York to this accuracy, it would be exact to the thickness of a human hair. That's how delicately quantum electrodynamics has, in the past fifty years, been checked—experimentally and theoretically.

—Richard Feynman[1]

THE ELEMENTS

As the eighteenth century wound to a close, the universe appeared to physicists, or "natural philosophers" as they were then known, to be composed of two kinds of stuff—matter and light. These interacted with one another by means of three forces—gravity, electricity, and magnetism. Thanks to the work of John Dalton and other chemists starting in the early nineteenth century, it was determined that all material bodies were composed of just ninety or so *chemical elements* that individually could not be broken down further, at least by the technology of the day. In 1869 these elements were systematically grouped by the Russian chemist Dmitri Mendeleev into the Periodic Table we still see today (in more up-to-date versions) on the wall of every chemistry classroom.

Familiar substances like air and water were found not to be elementary but could be broken down into combinations of these elements. Air is a mixture of (mainly) nitrogen and oxygen. Water is composed of molecules made from hydrogen and oxygen. Living matter comes in many forms but was found to always contain carbon, along with an array of other elements.

Although the matter of everyday experience, such as trees, water, and air, looks continuous to the naked eye, scientific observations in the late nineteenth century led to the proposal that material bodies are mostly empty space filled with tiny particles too small for us to see. At that time the simplest of these particles were the chemical elements that, for historical reasons, were called "atoms" following the term used by Leucippus and Democritus.

The term *atom* derives from *atomos*, or "uncuttable" in Greek. The groupings of atoms into substances such as H_2O (water) and NaCl (table salt) are called *molecules* and, while not pointlike particles, they are still small enough to be invisible to the naked eye.

Not all nineteenth-century scientists accepted the atomic view of matter. The eminent physicist and philosopher Ernst Mach believed that physics should only deal with observable phenomena, so to him talk of invisible "atoms" was pointless speculation. Most chemists viewed the atomic model as a handy tool and that, lacking any good evidence otherwise, assumed that matter was in fact a continuous fluid. It was not until early in the twentieth century that the existence of atoms was indirectly verified and it would take until the 1980s before individual atoms could be directly photographed. A lesson to be learned from this story is that science can often say much about that which cannot be directly observed by the technology of the day.

THE STRUCTURE OF ATOMS

In 1896 Antoine Becquerel discovered that uranium, the heaviest element in the Periodic Table as then known, produced invisible emanations that exposed photographic plates in a dark drawer. Pierre and Marie Curie extensively investigated this new phenomenon, called *radioactivity*, in the laboratory. Marie would get two Nobel Prizes for her lifetime's achievements. Her death in 1934 was probably the result of her long-term exposure to radiation that no one at the time suspected might be harmful.

In the early 1900s, Ernest Rutherford demonstrated that radioactivity resulted from the spontaneous disintegration of atoms. That is, the atom was not "uncuttable" after all. Instead, it was apparently composed of smaller, more basic constituents. The "elements" of the Periodic Table are not elementary.

In 1909 a team led by Rutherford bombarded gold foil with *alpha-rays* from radium, another highly radioactive element. From the fact that these rays scattered from the gold target at very large angles, Rutherford inferred that the atom was composed of a tiny nucleus surrounded by even tinier orbiting electrons. Later investigations showed that the alpha-rays themselves were actually nuclei of helium atoms.

QUANTUM THEORY OF THE ATOM

Rutherford had likened the structure of the atom to that of the solar system, which is mostly empty space with the sun orbited by much smaller planets. However, this analogy turned out to be too crude. Unlike the planets, electrons have electric charge and, according to electromagnetic theory, radiate electromagnetic waves when they change direction, as they do in orbit. According to the physics of the time, an electron in an atom should lose energy and quickly spiral into the nucleus. Obviously this does not happen and some new physics was needed.

In 1913 Niels Bohr introduced the "quantum theory" of the atom. Inspired by Planck‘s notion of light quanta, Bohr proposed that an electron in an atom can occupy only certain allowed orbits, specifically those in which the electron's angular momentum is an integral number of Planck's constant divided by 2π. When atoms are "excited," for example, by sending a high-voltage electric spark through a gas as in a fluorescent light, electrons are pumped up into higher orbits, or "energy levels." They then drop spontaneously from the higher orbits to lower ones, emitting quanta of radiation, that is, photons.

When it has dropped to its lowest energy level, an electron in an atom can drop no further and the atom remains stable thereafter. Bohr worked out his model in mathematical detail for the simplest atom, hydrogen, which is composed of a nucleus containing (in the most common form) a single proton with a single electron in orbit. He successfully described the spectrum of light emitted by atomic hydrogen, which occurs in discrete steps and had been a puzzle since first observed in the nineteenth century. As you move up in the Periodic Table you add an electron and a proton in each step. Because of the large number of particles, it quickly becomes impossible to calculate the atomic structure mathematically and approximation techniques must be used. Nevertheless, the quantum nature of atoms explains their complexity without which life would not exist, along with almost everything else of interest in the universe.

Quantum theory also introduced chance where previously we had deterministic cause. Nothing "causes" an electron in an excited atom to drop to a lower energy level at a given time. All the early quantum theory could calculate then, and all the most advanced quantum theory can calculate today, are the probabilities for atomic transitions and other phenomena such as nuclear decay. This fact refutes the claims of theistic philosophers since Aristotle, theologians since Thomas Aquinas, and contemporary Christian apologists such as William Lane Craig, that everything must have a cause and the prime cause is God.

THE STRUCTURE OF NUCLEI

The discovery of the *neutron* in 1932 by James Chadwick established that the atomic nucleus is composed of protons and neutrons. Their masses are almost the same, with the neutron about 0.1 percent heavier than the proton. Protons have a positive electric charge and so tend to repel each other. Only opposite charges attract. Neutrons are electrically neutral. So, clearly, they are not being held in the nucleus by the electric force. A simple calculation shows gravity is far too weak to hold a nucleus together so a third force, which physicists dubbed the *strong force*, was needed.

And so familiar matter was found to be composed of just three particles: the proton, the neutron, and the electron. Light was also found to be composed of material particles called photons. Thus, in 1932 these four particles seemed to be all we needed to describe the material universe. Whether a sample of matter was taken from the air, the ocean, a living organism, or a meteorite from space, when analyzed for its most basic constituents scientists found nothing but protons, neutrons, and electrons.

The problem was that other material objects kept showing up that were not simple combinations of protons, neutrons, and electrons. One such particle is the *neutrino.* In the form of nuclear radiation known as *beta-decay,* an electron is emitted by the nucleus. The basic process is a neutron decaying into a proton and an electron. However, when one adds up the energy of the proton and the electron, it is less than the rest energy of a neutron. Some energy is carried off without being seen.

THE NEUTRINO

This might have been interpreted as a violation of the fundamental principle of conservation of energy. However, in 1931 Wolfgang Pauli proposed that another particle having zero electric charge and extremely low—if not zero—mass was also being emitted. This very elusive particle, which Enrico Fermi dubbed the *neutrino* ("little neutral one"), was not detected directly until 1956 by Clyde Cowan and Frederick Reines. I spent a good half of my research career working with neutrinos, some of it in collaboration with Reines, who belatedly received the Nobel Prize in 1995, shortly before his death. As we will discuss later, there are three types of neutrinos: electron neutrinos, muon neutrinos, and tauon neutrinos. (See below for a discussion of muons and tauons.)

In my last project before retiring from research, I was involved in an experiment in Japan called *Super-Kamiokande,* which showed for the first time, in 1998, that neutrinos have nonzero mass. This experiment was based on an idea I first proposed in 1980.[2] Muon neutrinos are produced in great numbers by cosmic rays hitting the top of the atmosphere. They can be registered by large-volume detectors designed to look at neutrinos from the sun or other astronomical bodies. The detectors, such as Super-Kamiokande, are placed deep underground or undersea to filter out the background cosmic ray particles that penetrate from the surface above the detector.

The muon neutrinos produced on the other side of Earth from the detector pass through Earth, traveling twelve thousand kilometers before reaching the detector. If neutrinos have mass, then there is a chance for some of the muon neutrinos from the other side of Earth to "transmute" into neutrinos of one of the other two types—electron neutrinos or tauon neutrinos. This is unlikely to happen for the muon neutrinos coming straight down, which travel only about twenty kilometers. If neutrinos have mass there will be fewer muon neutrinos coming up through Earth than down from the sky. This is exactly what was observed.

The Japanese leader of the Super-Kamiokande experiment, Masatoshi Koshiba, shared the 2002 Noble Prize in physics for this work.

ANTIMATTER AND COSMIC RAYS

The simple picture of a small number of particles constituting all of nature did not live out the year 1932 when, as mentioned above, the neutron was first observed in the laboratory. In 1928, Paul Dirac developed a relativistic (that is, took into account Einstein's special theory of relativity) version of quantum mechanics that predicted the existence of an *antielectron*, a particle with the same mass and other properties as the electron except with opposite (positive) electric charge. The existence of the antielectron, or *positron*, was confirmed in 1932 in a cosmic-ray experiment by Carl Anderson. Twenty years later, antiprotons and antineutrons were produced on the campus of the University of California at Berkeley, by the Berkeley Bevatron.

Thus it was confirmed that a kind of mirror matter—*antimatter*—exists that does not occur naturally except in high-energy collisions such as those of cosmic rays with Earth's atmosphere. Today, the antimatter in the universe constitutes only one part in a billion of normal matter. At one time, however—very early in the life of our universe—there were likely equal amounts of matter and antimatter. When matter and antimatter meet, they annihilate, turning into photons. In the early universe, a slight difference in the strengths of the interactions of matter and antimatter resulted in almost all the antimatter disappearing into photons, leaving a small residue of matter—one part in a billion—to make up the universe. If this asymmetry between matter and antimatter did not exist, the universe today would be nothing but cold photons and neutrinos.

The photons produced in the annihilation have now cooled to just 2.7 degrees Celsius above absolute zero and fill the universe in the form of the cosmic microwave background radiation (CMBR) discovered in 1964 by Arno Penzias and Robert Wilson. This discovery provided direct evidence for the *big bang model* of the early universe that today is solidly confirmed by a wide range of astronomical data over the entire electromagnetic spectrum taken with telescopes on Earth and in space.

In 1937 Anderson and his collaborator Seth Neddermeyer saw a new particle in cosmic rays that seemed to be nothing more than a heavy electron, a particle now called the *muon.* "Who ordered that?" muttered physicist Isidore Rabi. The muon turned out to be identical to the electron, only 207 times heavier.

World War II brought to a halt any research not directly associated with the war effort. When basic research resumed after World War II, invigorated by a new crop of young physicists, a series of new particles were seen in cosmic rays and physicists began to wonder where they all fit into the scheme of things. Protons, neutrons, and electrons composed the matter that made up the planets and stars, rocks and trees, you and me. So who needed more elementary particles?

By the late 1950s we knew that the cosmic rays hitting Earth from outer space were mostly high-energy protons that were absorbed in collisions with atoms in the upper atmosphere. These collisions produced new particles, and only some of these reached Earth's surface, mostly neutrinos and muons. Neutrinos are so weakly interacting with other matter that they pass through our bodies (and the Earth) as though through empty space. Muons are electrically charged and so interact electromagnetically, but still sufficiently weakly that they also penetrate deep into matter—though not anywhere near as deep as neutrinos.

The most common particles produced by high-energy collisions of protons with matter are very light particles called pi mesons, or *pions.* These are very short lived, decaying into even lighter particles such as photons, neutrinos, muons, and electrons.

THE BIRTH OF PARTICLE PHYSICS

In the fall of 1959 I began graduate research in elementary particle physics at the University of California at Los Angeles (UCLA). The field was a new one, having evolved out of nuclear physics just a few years earlier. The experience of World War II, especially the nuclear bomb, had driven home to politicians and the public alike that neglecting fundamental physics carried with it dire implications for national security. As a result, money poured out of Washington to those universities and national laboratories equipped to carry out the needed research.

One of the best equipped was the University of California at Berkeley, whose physics department led the way in nuclear physics with its star-studded faculty and Lawrence Radiation Laboratory (later renamed Lawrence Berkeley Laboratory to rid it of the dreaded term "Radiation") in the beautiful hills above the main campus. In 1932 Ernest Lawrence had built the first cyclotron at Berkeley, which accelerated protons to an energy of 80,000 electron-volts.[3] Somewhat later, Robert Oppenheimer was brought in from Berkeley to direct the Manhattan Project and build the Bomb. Now, in 1959, a new particle accelerator called the *Bevatron*, was in operation on the hill.

The Bevatron accelerated protons to an unprecedented six billion electron-volts (hence Bev-atron). In 1955 it produced the first antiprotons observed in the laboratory. Actually, this was not a surprise. As seen above, the antielectron or positron had been predicted by Dirac in 1928 and observed by Anderson in cosmic rays as far back as 1932. By the 1950s, physicists knew that producing antiprotons was just a matter of colliding two protons together with sufficient energy. On my PhD oral exam I was asked to calculate that threshold energy; I can still do it in two lines. The answer is six billion electron-volts, not coincidently the Bevatron's design energy.

By the time I began my research with data from the Bevatron, cosmic-ray studies had revealed the existence of other particles that were dubbed "strange particles" because of their odd behavior. The controlled beams produced by particle accelerators made it much easier to study such particles and unravel the source of their strangeness. The Bevatron actually produced a beam of strange particles called K-mesons, or *kaons*, that I would use for my PhD thesis research.

By shear luck I had stumbled into a field that was on the threshold of one of the greatest scientific breakthroughs in history: the discovery of the fundamental objects and forces that provide the underlying framework for the universe in which we live. When I finished my degree in 1963 and accepted an assistant professorship at the University of Hawaii, dramatic developments had already taken place.

Other laboratories in the United States and Europe joined the Bevatron in producing beams of high-energy particles. Governments, again more for reasons of national security than any sincere striving for fundamental truth, generously funded these efforts—spreading the largess around, even to places as remote as the new state of Hawaii, where the state government, with federal help, built a first-class research university.

My first several years in Hawaii were spent—with Hawaii colleagues Vince Peterson (group leader) and Bob Cence—collaborating with a research group in Berkeley. I often visited and spent enjoyable summers in the hills overlooking San Francisco Bay.

THE PARTICLE EXPLOSION

Most of the new particles that were observed in cosmic rays and accelerator experiments proved to be unstable, disintegrating into lighter particles in small fractions of a second. Some survived no longer than a trillion-trillionth of a second, traveling not much more than the diameter of a proton (10^{-15} meter) before decaying. These extremely short-lived particles left no detectable signal in any device, but were inferred from the missing energy and momentum in the observable products of the high-energy collisions. As at the end of the nineteenth century, physicists were again finding evidence for components of the universe that were not directly observable, but by now no one questioned the validity of this process.

Despite not knowing what to do with all these new particles, physicists continued to accumulate them. In 1959 they numbered about ten separate varieties. By the time I completed my PhD four years later, about a hundred new varieties of particles had been reported.

While my fellow experimentalists and I were busily gathering data, theorists were seeking to uncover some underlying order in what seemed to be a great confusion of objects. Obviously all these particles could not be elementary. Indirect evidence already indicated that the proton and the neutron were not "elementary," that is, that they possessed a substructure. The obvious path was the one that had worked well so far: search for the "atoms," the basic objects from which the observed particles themselves might be constructed.

Now, you might ask, why not just collide protons with other particles, such as photons or pions, to look at what is inside? This would simply mimic Rutherford's 1907 experiment that discovered the atomic nucleus. Indeed, we were doing just that with the new high-energy accelerators. Unfortunately, bombarding protons in a fixed target with high-energy protons or pions only results in more particles—mostly pions, with a few kaons and small amounts of about everything else for which there is sufficient energy to produce including the whole catalog of new, unexpected, unasked-for particles. This was the direct result of $E = mc^2$. In the nuclear bomb this relationship is exploited to convert rest energy into kinetic energy. In the particle collisions this was turned around, with collision kinetic energy being converted into rest energy and ultimately different forms of mass.

The great bulk of the particles being discovered at accelerators were formed from the collisions of beams of protons shot at targets of protons and neutrons, which are governed by the strong nuclear force. These particles were dubbed *hadrons*.

FOUR FORCES

Besides observing new particles, elementary particle physicists were concerned with how these particles interacted with one another. By 1959 it was understood that there had to be at least four fundamental forces in nature.

1. Gravity. All material bodies attract one another by gravity. In 1916, Einstein's general theory of relativity replaced Newton's law of gravity as the basic theory of gravity, although Newtonian gravity still works fine for most purposes, such as calculating the orbits of large space vehicles. Truly precise calculations based on orbital mechanics, such as calculating an Earth-bound location to a precision of a few meters via GPS satellites, require use of Einstein's theory. To this day, the general theory remains consistent with the most precise and elegant experiments that have been devised to test it. However, gravity is far weaker than the other forces that act at the elementary particle level and until recently has been generally neglected in these studies.

2. Electromagnetism. In the mid-nineteenth century physicists realized that the electric and magnetic forces were different aspects of a single, unified force. As classical fields, they are well described by Maxwell's equations, first presented to the Royal Society by James Clerk Maxwell in 1864. At the quantum level, a successful quantum field theory of electromagnetism called *quantum electrodynamics* (QED) was developed in the 1940s by Richard Feynman, Julian Schwinger, Sin-Itiro Tomonaga, and Freeman Dyson.[4]

The basic interaction between electrons in QED is illustrated in figure

9.1 (a), a pictorial representation that is called a *Feynman diagram.* An electron emits a photon that travels across space where it is absorbed by another electron, carrying energy and momentum from one electron to the other in the process. The fact that the photon has zero mass implies that the range of the interaction is limitless, which makes it possible for an electron in a galaxy a billion light-years away to interact with an electron in the detector of a telescope on Earth.

Figure 9.1 (a) is just the "lowest-order" Feynman diagram. In principle there are an infinite number in which all kinds of internal processes take place. For example, figure 9.1 (b) shows a higher-order process in which one of the electrons emits a photon (dashed line) that is absorbed by another electron. Figure 9.1 (c) shows another possibility, where the exchanged photon transforms into an electron-positron pair, which then annihilate back into a photon.

Fortunately, in the case of QED, the electromagnetic force is sufficiently weak so that an approximation technique called *perturbation theory* makes it possible to make an accurate calculation with just a few Feynman diagrams, all but a few of the higher-order diagrams becoming negligible.

It should be noted that Schwinger did not use Feynman diagrams in his QED calculations. Rather, he developed mathematical techniques of such complexity that few besides himself and a few devoted students could carry them out. Even his students reverted to the simpler Feynman technique when the boss was not looking.

Still, Schwinger achieved the first great triumphs of QED. In 1948 he precisely calculated the magnetic strength of an electron ("magnetic dipole moment"). The following year Schwinger ensured his trip to Stockholm to receive a Nobel Prize by computing the tiny shift in specific spectral lines of hydrogen called the *Lamb shift.* Both calculations agreed beautifully with high-precision experiments that had shortly before been performed and had been unexplained by previous theories.[5] As of this writing, QED has been tested to one part in a trillion or better. The term *uncertainty* is associated with quantum theories because of the uncertainty principle. This does not mean quantum theories, as we see in the case of QED, themselves are imprecise.

3. Weak nuclear force. This force operates only at tiny distances much smaller than a nuclear diameter. It is responsible for the *beta-decay* of nuclei and provides the main source of energy at the center of stars such

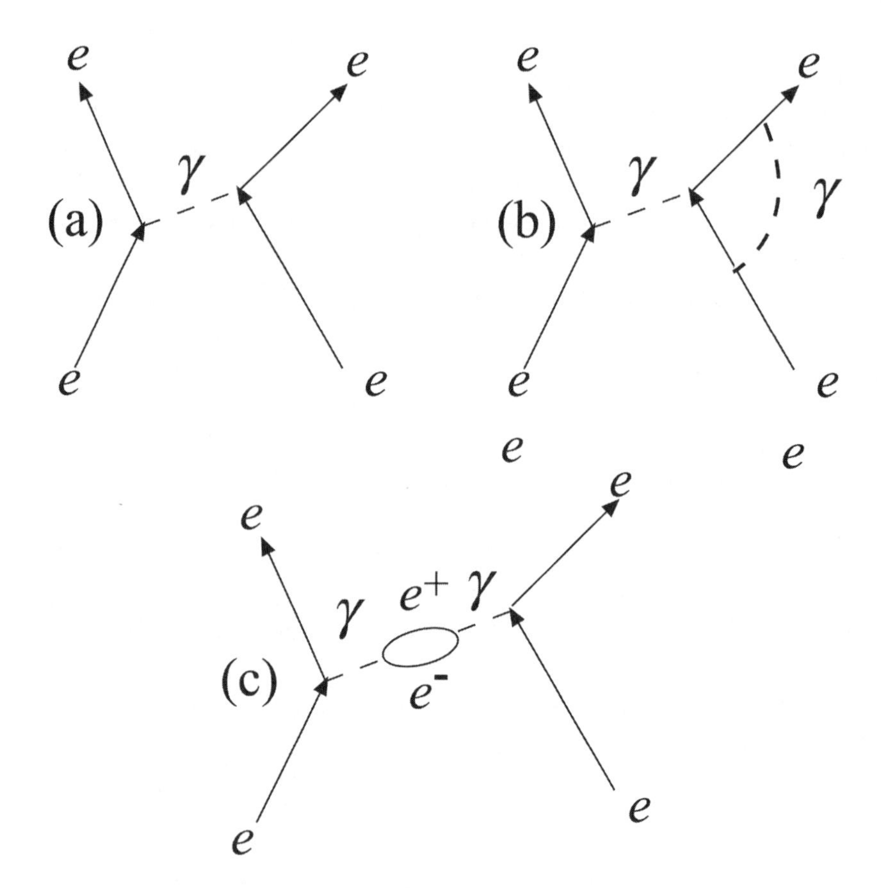

Fig. 9.1. Three examples of QED Feynman diagrams.

as our sun. The weak force is visualized by the Feynman diagram shown in figure 9.2 in which a negatively charged, hypothetical *W*-boson is emitted by a neutron (in or outside a nucleus), changing the neutron to a proton. The W^- then decays into an electron, *e* and an electron antineutrino, $\bar{\nu}_e$. (As we will see, there are three types of neutrinos and antineutrinos).

The range of a force mediated by an exchanged particle is inversely proportional to the mass of the particle. Since the range of the weak force is tiny, much less than the size of a nucleus, the *W*-boson must be massive—many times the mass of a proton.

4. Strong nuclear force. Since the discovery of the nucleus it has been recognized that a special force of great strength and short distance

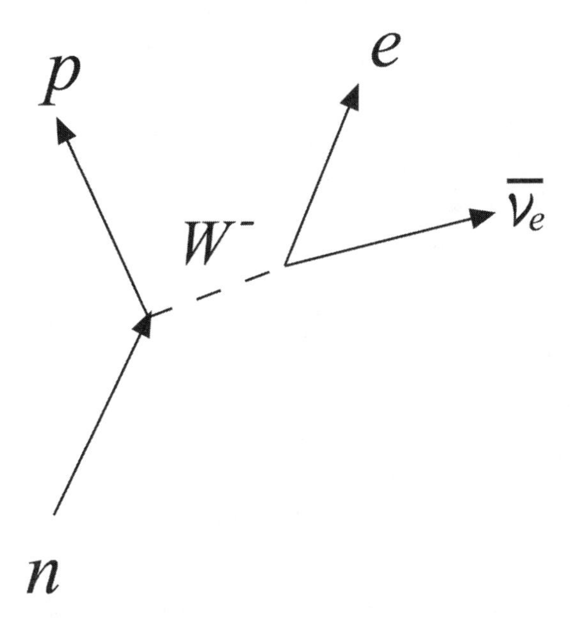

Fig. 9.2. Feynman diagram for neutron decay.

must act to hold the protons and the neutrons together in the nucleus. In 1935 the Japanese physicist Hideki Yukawa had proposed that protons and neutrons interacted by means of the exchange of particles not yet seen at that time called *mesons*, as illustrated in figure 9.3. In order for the range of the interaction to be about the diameter of a nucleus and not much bigger, he estimated that the rest energy of the meson must be about 140 million electron-volts (MeV).[6] This is to be compared with the electron, whose rest energy is 0.511 MeV and the proton whose rest energy is 938 MeV. Thus the meson was intermediate in rest energy between the two, hence the name *meson*. (Recall rest energy = $E = mc^2$, where m is the mass.)

When a particle was discovered just two years later, in 1937, and its rest energy measured to be 106 MeV, it seemed that Yukawa had triumphed. But then physicists asked, how could a particle responsible for the strong force between protons and neutrons pass through the whole thickness of the atmosphere and even deep into mines after being produced by cosmic rays at the top of the atmosphere?

The new particle was not Yukawa's meson but a heavy version of the electron we now call the *muon*. But eventually, after World War II, the

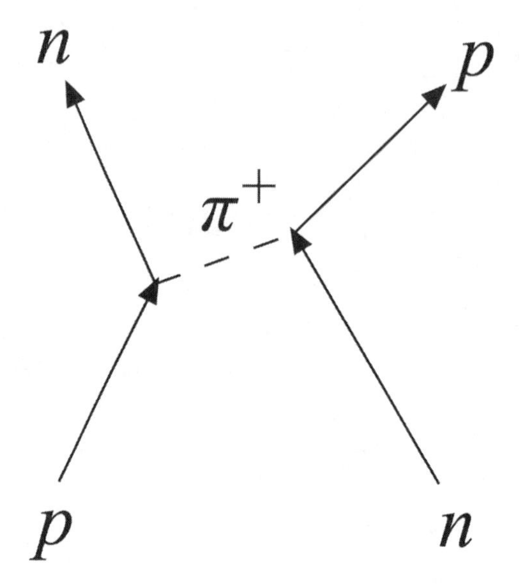

Fig. 9.3. Example of proton-neutron interaction with charged pion exchange.

Yukawa particle was seen in cosmic rays at the top of a mountain and its rest energy measured to be 140 MeV, exactly as predicted. Since other particles were later discovered that were generically called mesons, Yukawa's particle is now dubbed the *pi meson*, or *pion*.

However, despite the remarkable success of Yukawa's prediction, attempts to fit pion-exchange models to proton-proton and proton-neutron scattering achieved, at best, only crude, qualitative success.

By 1959, my first year in physics research, no successful theory of either the weak or the strong force had yet been developed. The attempts being made at this time to understand the strong force by a new type of "holistic" theory will now become an important part of our story.

PROBLEMS WITH FIELD THEORY

The success of QED made it natural to attempt to develop a theory of the weak and the strong nuclear forces along the same line. However, despite its great computational accuracy in certain problems, QED only worked because of two lucky circumstances. QED is a quantum field theory in which particles are the "quanta" of continuous fields in space and time. For example, the photon is the quantum of the electromagnetic field. The electron is the quantum of the Dirac field.

Continuous quantum fields are plagued by a notorious mathematical problem. Calculations made with these fields tend to go to infinity at small distances. QED succeeded when a mathematical trick called *renormalization* was discovered that enabled these infinities to be subtracted away.

That was the first lucky circumstance. The second was that perturbation theory, as described earlier, enabled physicists to carry out practical calculations. They were immensely difficult, but they were doable. When theorists tried to apply these same methods to the weak and the strong nuclear forces, they turned out not to be able to achieve any useful solutions.

The *W*-boson that was hypothesized to mediate the weak interaction and the pion that was hypothesized to mediate the strong interaction are unlike the photon in two ways: (1) they are massive and (2) they are electrically charged (although a neutral pion was also known at this time and a neutral companion to the *W*-boson was later found). Furthermore, the strong force is so strong that perturbation theory fails; each higher-order diagram is more important than the previous lower one!

At this point many of the best minds in the field, notably the brilliant Russian physicist Lev Landau, were ready to give up on quantum field theory as "fundamentally, logically incomplete."[7] Even QED, they found, was bound to break down at high energies and so was not a complete theory.

THE BOOTSTRAP

The effort to find an alternative to field theory was led by a tall Berkeley professor with movie star looks named Geoffrey Chew. Chew and his followers reasoned that quantum fields were not directly measurable and therefore unphysical and meaningless. This was reminiscent of Ernst Mach's position with respect to atoms in the nineteenth century. Mach said he would not believe in anything he could not see. Today we can routinely view atoms with a *scanning tunneling microscope.*

Mach was a proponent of the philosophical doctrine termed *positivism.* A variation called *logical positivism* formed a serious philosophical school of thought early in the twentieth century (see chapter 13). The idea was to express all knowledge in terms of a language limited to observable phenomena. Nothing was real unless it could be observed, at least indirectly. One of the tenets of logical positivism was that metaphysics, which deals with the unobservable, is nonsense. Chew and his cohorts viewed quantum fields as metaphysical nonsense.

Chew not only envisaged doing away with quantum fields, he also argued that there were no elementary particles or, more precisely, that all particles were equally elementary—being composed in some sense of one another. Quantum mechanics had shown that the state of a system can be expressed as a combination of all its possible configurations. So, Chew reasoned, a pion is sometimes a proton-antiproton bound state, while a proton is sometimes a neutron and a positive pion, or a sigma-plus particle and a neutral kaon, and so on. Chew called this "nuclear democracy" and the theory appeared under various names such as *bootstrap theory* or, more formally, *S-matrix theory*.

Chew had also argued that the traditional space-time description of physical events was not fundamental and so need not be a part of fundamental theory. He pointed out that the basic process taking place in nature was the scattering of particles from one another, and this is best described in terms of *momentum* and *energy* rather than *position* and *time*. The experiments we were doing, after all, involved the collisions of particles from an accelerator, usually protons in those days, with some target nucleus, which also contained protons and neutrons. We do not trace the particles' positions in space and time. Indeed, at the heart of the reaction, the events occur in an unobservably small region of space. Rather, we set up the experiments so the beam and the target's momentum and energy are known. Then we measure the momenta and energies of the particles that come out after the collision. The probability that particular reaction will occur is given in terms of a quantity called the *S-matrix*, defined in 1941 by Werner Heisenberg, the author of the Heisenberg uncertainty principle. It does not depend on space and time but on momentum and energy.

Chew proposed that the S-matrix, written as a function of energy and momentum, must conform to certain mathematical principles ("symmetry," "unitarity," and "analyticity") that are too technical to discuss here. Suffice it to say, it was conjectured that from these principles alone, all the particles and their properties could be determined. We will see how well this worked later. In the meantime, we need to reintroduce the person who, more than any other, influenced the quantum spirituality movement—physicist Fritjof Capra.

THE TAO AND THE BOOTSTRAP

In *The Tao of Physics*, which we discussed in chapter 3, Capra described bootstrap theory, as he gleaned the principles from Chew's writings, as follows:

> The bootstrap philosophy constitutes the final rejection of the mechanistic world view in modern physics. Newton's universe was constructed from a set of basic entities with certain fundamental properties, which had been created by God and thus were not amenable to further analysis. In one way or another, this notion was implicit in all theories of natural science until the bootstrap hypothesis stated explicitly that the world cannot be understood as an assemblage of entities which cannot be analyzed further. In the new world view, the universe is seen as a dynamic web of interrelated events. None of the properties of any part of this web is fundamental; they all follow from the properties of the other parts, and the overall consistency of their mutual interrelationships determines the structure of the entire web.[8]

To Capra, bootstrap theory was a beautiful example of the connectedness that he found in Eastern philosophy. He further states, "The notion of elementary particles as the primary units of matter has to be abandoned."[9]

After having done research elsewhere in the San Francisco Bay area, Capra joined Chew's group in 1975, just having written his best seller and, as we will see, just as the bootstrap began to unravel.

NOTES

1. Richard P. Feynman, *The Strange Theory of Light and Matter*, paperback ed. (Princeton, NJ: Princeton University Press, 1988), p. 7.

2. V. J. Stenger, "Neutrino Oscillations in DUMAND," *Proceedings of the Neutrino Mass Miniconference, Telemark Wisconsin*, October 2–4, 1980, University of Wisconsin report 186, 1980, pp. 174–76.

3. The electron-volt (eV) is the energy an electron (or a proton) gains when falling through an electric potential of one volt.

4. Sylvan S. Schweber, *QED and the Men Who Made It* (Princeton, NJ: Princeton University Press, 1976). Dyson is not usually given his due recognition, which is an unfortunate consequence of the fact that Nobel Prizes are limited to three recipients.

5. Ibid., ch. 7; see also Victor J. Stenger, *Timeless Reality* (Amherst, NY: Prometheus Books, 2000), ch. 7, for a less technical review of QED.

6. Since $E = mc^2$, mass units can be expressed as energy units divided by c^2.

7. David Gross, "Asymptotic Freedom and the Emergence of QCD," in *The Rise of the Standard Model: Particle Physics in the 1960s and 1970s*, ed. Lillian Hoddeson, Laurie Brown, Michael Riordan, and Max Dresden (Cambridge: Cambridge University Press, 1997), pp. 202–203.

8. Ibid., p. 286.

9. Ibid., p. 75.

10

CHAOS, COMPLEXITY, AND EMERGENCE

> **Chaos begets complexity, and complexity begets life. Without chaos, we would not be here.**
>
> —John Gribbin[1]

NEW PHYSICS AT THE MACROSCALE

By the nineteenth century the mathematical methods of Newtonian mechanics had been developed sufficiently so that many everyday phenomena could be handled. Not all could be solved exactly. In fact, it was shown that only systems of one or two bodies were solvable in principle. Three and more requires approximations.

Approximation techniques such as *perturbation theory* enable physicists to make useful calculations for systems of a few bodies that interact with one another weakly. This works well for the planets and other bodies in the solar system. But this stratagem fails when the bodies strongly interact.

Of course, it is hopeless to calculate the detailed motion of the trillion-trillion atoms in a gram of familiar matter. Instead, physicists use statistical techniques to calculate the average behavior of such systems. With the methods of *statistical mechanics*, also developed in the nineteenth century, many of the gross properties of the gases, liquids, and solids of normal and laboratory experience can be computed. However, these systems must be in or not too far from thermal equilibrium (constant temperature throughout). This method fails for many multibody systems that are far from equilibrium, such as Earth's atmosphere with its turbulence and strong interactions with land and sea.

With the development of computers in the twentieth century, it became possible to explore by simulation the behavior of multibody systems that could not be handled by standard mathematical techniques. The motion of three or more bodies became predictable, depending only on computer power. And physicists were no longer limited to weakly interacting systems such as the planets. They could take on more-complicated systems in which the particles interacted strongly.

The results were unexpected and significant. New phenomena were discovered on the macroscale and some thought they represented another paradigm shift in our understanding of the universe. One such phenomenon was called *chaos*.

DETERMINISTIC CHAOS

Chaos was accidentally discovered in 1961 by meteorologist Edward Lorenz while running a model of the atmosphere on one of the primitive computers of the day. In repeating a run, he entered a number that had been rounded off to 0.506 in a printout from the actual number inside the computer, which was 0.506127. He found that the model gave completely different results in the two cases. Lorenz discovered that his model was very sensitive to tiny changes in the input data. It was as if a butterfly flapping its wings could change the weather days ahead, so this was dubbed the *butterfly effect*.[2]

Computer simulations discovered that the butterfly effect and other unexpected phenomena are associated with systems that have three basic characteristics:

1. *Nonlinearity*. A linear system is one whose output response to a stimulus is proportional to the stimulus. For nonlinear systems this is not the case.

2. *Energy dissipation.* The system must have a means of losing energy, such as friction.
3. *External driving force.* An outside force must act on the system. This also provides some or all of the energy lost to dissipation.

These systems appear to behave unpredictably and so the general phenomenon was given the misleading name "chaos."

A simple example of a system that meets these characteristics is the damped, driven pendulum. A pendulum will respond linearly to a slight push, but its response becomes nonlinear as the push gets harder. Add damping and the pendulum behaves chaotically.

No known mathematical technique enables one to go from the initial conditions to the final results of a chaotic system, and so the process appears on the surface to be indeterministic. However, for a system on the macroscale, an individual body such as a pendulum bob still obeys deterministic Newtonian mechanics.

The apparent unpredictability of a chaotic system is the result of our own limited knowledge of the initial conditions. When we do computer simulations on chaotic systems we can predict the outcome even if we can't calculate it by traditional mathematical means such as mathematically calculating the laws of motion. All we need to do is run the simulation once and see where the system ends up. Then, as long as we run it again on the same computer from the same initial point (taking care to avoid rounding errors), we will end up at the same final point. For these reasons, the chaos associated with nonlinear systems is more accurately denoted as *deterministic chaos.*

Quantum systems, which are only statistically deterministic, are linear and so do not exhibit this variety of chaos. Attempts to develop a nonlinear version of quantum mechanics have so far failed. In fact, linearity is behind many quantum effects. Nonlinear quantum mechanics would have none of what Einstein called the "spookiness" that makes it so weird in the first place and so attractive to those looking for a scientific basis for their own weird beliefs.

While deterministic chaos is limited to classical systems, quantum uncertainties in the initial conditions could result in a large-scale, otherwise deterministic chaotic system such as a pendulum or the atmosphere, to behave unpredictably. Note that this is not "quantum chaos" since once the initial conditions are set the system still behaves deterministically.

SELF-SIMILARITY AND SELF-ORGANIZATION

Chaotic systems were found to exhibit many other interesting and unanticipated features that cannot be derived from fundamental physics. One of the most fascinating features of chaotic systems is *fractal* behavior, whereby the system undergoes certain patterns of motion that repeat themselves as one goes to smaller and finer detail. This property is called *self-similarity.* Computer-generated fractals have produced images that strongly resemble natural objects such as clouds, snowflakes, coastlines, and mountain ranges. However, these physical objects do not possess self-similarity and so are technically not fractals. A simple Internet search will direct the reader to countless examples of fractals.

Some chaotic systems exhibit a property of *self-organization* in which the simple can become complex without any conscious design or creative actions taking place.[3] For many examples of self-organization in both nature and computer simulations, see the beautifully illustrated book *The Self-Made Tapestry* by Philip Ball.[4]

Some have taken the remarkable properties of chaos to mean that a grand new holistic paradigm has been discovered in physics that will undercut and supplant the old reductionist methods with a new "science of wholeness" that fits in very well with the ideas of Capra and the new spiritualists.[5] However, decades have passed since the original excitement of the discovery of chaos and nothing so world-shaking as a new paradigm has taken place. If reductionist physics did not anticipate chaos, physicists have not found it necessary to introduce anything beyond reductionist physics to understand chaos.

COMPLEXITY

Related to chaos is the study of complexity. The two are often linked together as if they were a single subject, "chaos and complexity," but this is misleading. Chaos deals specifically with nonlinear, dissipative systems with an external driving force. These systems can be complex but also as simple as a pendulum. Complexity deals with many body systems that are organized in interesting ways. Some of those systems may be chaotic, but not all are.

The main message about complexity that I want to bring up for the purposes of this book is the fact that *complexity can arise naturally from simplicity.* This is one of those counterintuitive facts of nature that most people find difficult to believe and makes them sympathetic to those creationists who argue that the world, because it is complex, cannot have come about without divine intervention.

The development of complex systems from simpler systems has been demonstrated in virtually every field of science and, indeed, everyday life. Snowflakes develop spontaneously from water vapor. Amino acids and other molecules of life are easily assembled from basic chemical elements, although the origin of life itself is clearly not so simple. Once life exists, organisms develop from the splitting or merging of cells. As Ball has shown in his other admirable book *Critical Mass*, social systems such as markets, traffic, and international relations also exhibit spontaneous complex behavior that grows out of the simple interactions of their basic elements.[6]

The easiest way to see simplicity generate complexity in action is by means of *cellular automata.* This is best done on a computer, but let me describe it as a game played with a piece of graph paper in which you fill in squares with a pencil depending on some rule. Each square on the graph paper is called a "cell." If a cell is filled in it is "on," if not, it is "off." You start out with a particular pattern of on and off cells. Call that time $T = 0$. Then you go through a sequence of steps, $T = 1, 2, 3, \ldots$, where at each step you apply some simple rule that decides whether or not each cell remains on or off.

Perhaps the most remarkable cellular rule was that invented in 1970 by a young mathematician named John Hornton Conway for what he called the game of "Life." For an excellent discussion of Life including some deep implications, see *The Recursive Universe* by William Poundstone.[7]

Rules for the Game of Life:

Each cell has eight neighboring cells, four on its edges, and four on its corners. If the number of neighbors on is exactly two, the cell maintains the status quo, being on or off depending on whether it currently is on or off. If the number of neighbors on is exactly three, the cell will be turned on. Otherwise the cell is turned off.

That's all there is to Life. Conway called it "Life" because it roughly simulates growth patterns for simple living organisms such as bacteria. A cell with fewer than two neighbors dies of isolation. A cell with more than four neighbors dies of overpopulation. Two or three neighbors are just right.

Notice the player makes no decision in the course of play except to decide what cells are on at the outset. Otherwise she just follows the rules.

I do not have the space to illustrate all the remarkable automata that result just from different starting points and refer you to Poundstone. Better, you can play the game yourself. A quick search of the World Wide Web will turn up many sites that enable you to play cellular automata games online, including Life.

Start with three cells in a row and in two steps you get a "blinker" that flashes on and off. Four cells in a row lead to a "beehive," a two-by-three hexagon that is a "still-life," that is, changes no further. Other stable patterns that result from eight starting cells or fewer include, "ships," "aircraft carriers," "snakes," and "canoes."

Most patterns, however, are not still-lifes but change from step to step. A few are oscillators that, like the "blinker," blink on and off. These include "beacons" and "clocks."

Perhaps the most interesting and common of the simple automatons is the "glider," which is illustrated in figure 10.1. The glider starts with a pattern of five cells that repeats itself every four steps.

This is one you especially want to view on a computer, watching the glider move across the screen. Other moving patterns also exist and are generically referred to as "spaceships." Spaceships in general throw off "sparks" as they move along suggesting rocket exhaust.

Another interesting pattern is the "eater," which is a still-life that can

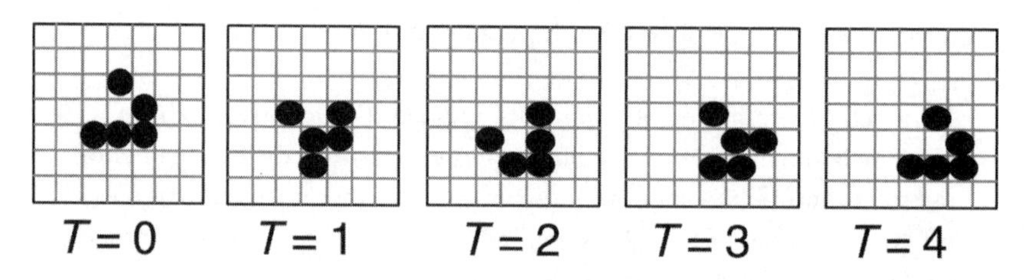

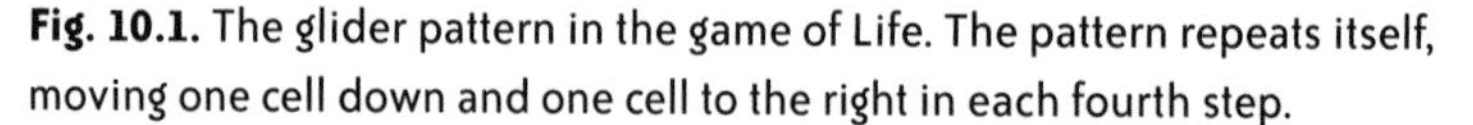

Fig. 10.1. The glider pattern in the game of Life. The pattern repeats itself, moving one cell down and one cell to the right in each fourth step.

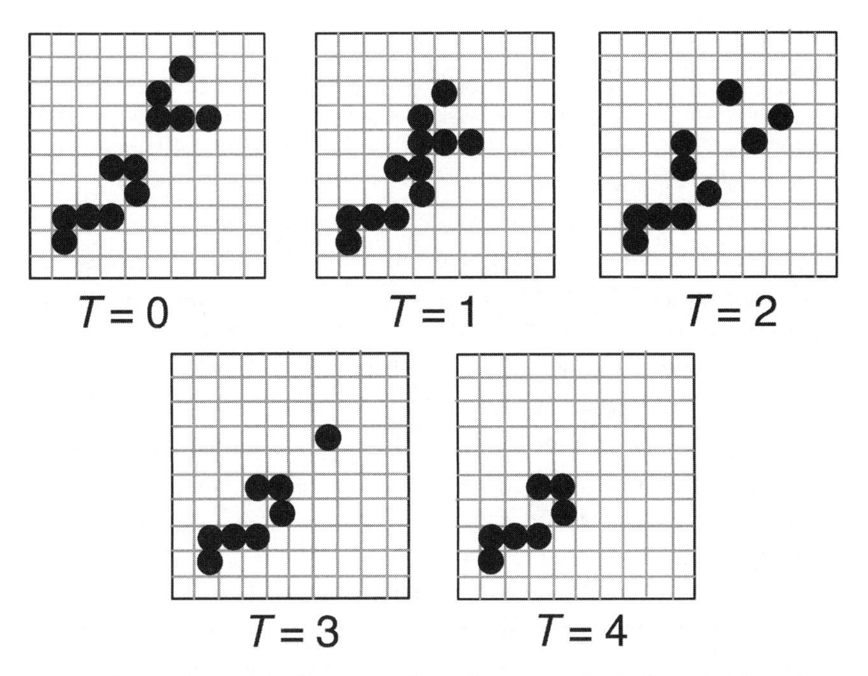

Fig. 10.2. An eater digesting a glider. The eater, a hook-shaped pattern, is stationary. The glider comes in from the upper right.

swallow up other objects that move its way. Figure 10.2 shows an eater digesting a glider.

When two gliders collide along a certain line they will annihilate into nothing. If they collide at an angle, they can produce a stable two-by-two cell "block." Or they can form a stable circular "pond" of eight cells. Spaceships show similar effects.

It is important to realize that none of these patterns were designed into the game. Conway had no way of knowing ahead of time that they would occur.

There is much more to the story of Life, the computer game, including the immensely profound fact that the algorithm itself can be used to create a universal computer.[8] In more recent years, Stephen Wolfram has studied a vast range of cellular automata and written a tome in which he makes the ambitious claim that the methods of cellular automata constitute a "new kind of science." He purports this to be a novel scientific method that is different from the traditional scientific method based on mathematics and that this new method can be applied in many fields.[9] These applications have not yet appeared.

In any case, I think I have sufficiently illustrated my point that simplicity can beget complexity. It follows that the complexity of the universe is no signal of intelligent design.

Theists often assert that "you can't get something from nothing." Without getting into the philosophical problem of defining "nothing" now (I will later), a random pattern of cells might be called "nothing" since it has no structure, and to be "something" requires structure. In any case, if you start a game of Life with a random pattern of off and on cells, gliders form naturally out of the chaos.

Of course, we still have the rules. But they are simple and can themselves be described as strings of on-off cells, or "bits," as can any logical statement. They can start off as "nothing" also—a random string of cells. Most would result in no interesting patterns. But one can easily imagine a natural process in which if one out of a random set of rules produced patterns that dominate over all the rest, generating "eaters" that gobble up everything but those of their own tribe. "Life" just might naturally evolve in the random universe.

EMERGENCE

The reductionist doctrine holds that matter can be broken down into parts that have their own independent identity. The simplest parts are elementary objects that cannot be broken down further (the Greeks called them *atoms*). According to our best current knowledge, these elementary objects are the quarks, leptons, and bosons of the standard model of particles and fields. Sometime in the early twenty-first century these may be found to have even more elementary constituents. We can only wait and see.

Despite its immense success, elementary particle theory has virtually no applications outside its own domain. Condensed matter physicists, those who study the properties of complex forms of matter such as liquid crystals or solid-state lasers, never bother to learn any elementary particle theory. They develop their own theories, based on nonrelativistic physics, to describe what they measure in their own "low-energy" laboratories. Similarly, chemists, biologists, geophysicists, meteorologists, and oceanographers currently have no use for relativistic quantum fields, quarks, and neutrinos. Yet most of these investigators would agree that the subjects of their interest are made up of elementary particles and probably nothing else. And we cannot rule out the possibility that future calculations on, say, global warming may require some input from relativistic quantum field theory.

Water is composed of two hydrogen atoms plus one oxygen atom. It is wet, a universal solvent, transparent to light, and so on. These are all emergent properties that do not describe atoms themselves. Who ever heard of a "wet oxygen atom"? But these properties follow from the properties of hydrogen and oxygen and the principles of atomic physics.

The new principles and properties discovered in the physical and biological sciences are completely consistent with basic physics. In all cases we find that the laws of physics are still obeyed: energy and momentum are conserved, forces are needed to accelerate objects. An eight-hundred-pound gorilla will fall from a tree with the same acceleration as an apple.

While a physicist will approximate the gorilla by a sphere, the zoologist needs more detail. The principles developed in zoology and the other fields that concern themselves with more-complex objects than particles and spheres are still material principles. We can imagine matter existing in a series of distinct levels of increasing complexity, from particles to atoms to molecules to inanimate bulk matter to animate matter, to thinking matter (brains), and beyond to human behavior and social systems. At each level a set of principles are said to "emerge" from the basic principles that lie beneath them all.

Emergence has become a hot topic in philosophy and theology. Many find the notion that we are "nothing but quarks and electrons" repellent and seek ways to show that we are "something more." These divide into two camps:

1. ***Material emergence***—those who maintain that the "something more" has evolved naturally from purely material systems and has no need to include God or anything supernatural.
2. ***Spiritual emergence***—those who see the emergence of "something more" as opening up a place for divine action.

Emergence has a long history that is unnecessary to review since that would add little to our discussion, whereas recent ideas are more relevant. Two scientists are representative of the *material emergence* camp.

Physicist and prolific writer Paul Davies has long argued that there must be organizing principles in nature over and above the known laws of physics that are yet to be discovered.[10] The principle that Davies sees emerging in complex systems is *self-organization.*

Biologist Stuart Kauffman has been the foremost proponent of self-organization in biology, claiming that chance and natural selection are insufficient for evolution. In his view, the natural tendency of material systems for self-organization also contributes to the development of biological complexity and may be necessary to explain the origin of life.[11]

Kauffman has recently published an ambitious book called *Reinventing the Sacred: A New View of Science, Reason, and Religion* in which he argues that no god is necessary. "The qualities of divinity that we hold sacred—creativity, meaning, purposeful action—are in fact properties of the universe that can be investigated scientifically."[12]

Representing the spiritual emergence camp is philosopher and theologian Philip Clayton, author of the definitive philosophical treatise on emergence, *Mind and Emergence: From Quantum to Consciousness.*[13]

All three authors largely agree on much of the subject, the main difference being that Clayton sees a possible opening for God while Kauffman wants to define a new religion in which "god" is inserted into a cold, lifeless universe. Davies has been sufficiently fuzzy about God in his writings to win the 1995 million-dollar Templeton Prize for Progress in Religion. Still he seems clear that emergence is purely material.

In this chapter I will limit myself to the discussion of emergence as a scientific phenomenon and forgo any religious or theological implications until chapter 15.

BOTTOMS-UP

If emergence simply means that particle physicists are incapable of deriving the principles developed by workers in most areas of great complexity, where that complexity is measured in terms of the large number of particles involved and their detailed interactions, then there is little more to say. If the conclusion is that the reduction of all material systems to elementary particles is thereby refuted and some additional laws of nature is required, then there is a lot to say. As I mentioned above, no laws of physics are violated in complex systems. Only when such a violation is observed or evidence is found that something more is involved, can materialist reductionism be ruled out.

Let us consider the first step in complexity, when we move from describing a single particle to two or more particles, say, electrons in quantum mechanics. Philosopher Paul Humphries points out that "quantum entanglement," in which a composite system can be in a pure state when the components of the system are not, leads to directly observable macroscopic phenomena such as *superfluidity* and *superconductivity*.[14] A more familiar example he does not mention is the Periodic Table of the chemical elements, which would not exist without the *Pauli exclusion principle*, which allows only one electron to be in a given quantum state at any specific point in space. This is a property that occurs only for a particle of half-integer spin (*intrinsic angular momentum*). You can have all the integer-spin particles such as photons you want in the same state. The complexity of chemical atoms, without which life as we know it would not exist, results from electrons filling various "shells" as you move up the table.

The Pauli principle naturally emerges when more than one electron is involved, but it is still derived from basic quantum mechanics. It is an example of emergence that is reducible to basic physics, what might be called *reductive emergence*. Emergence and reductionism are not incompatible.

Sorry for all these different types of emergence, but these definitions help clarify our meanings. Basically we have reductive versus *holistic* emergence and materialist versus spiritualist emergence. I will maintain that emergence is both reductive and materialist.

We find another example of reductive emergence when we step from individual particles to the situation where we have the trillions on trillions of particles in the bodies of everyday experience. In the nineteenth century, before the atomic theory of matter was established, physicists developed a set of principles to describe the behavior of material systems that enabled engineers to build heat engines and other devices involving heat transfer. As we recall from chapter 5, these principles included the first and second laws of thermodynamics. Once the atomic theory of matter was realized, these principles were derived from basic Newtonian particle mechanics. They are again examples of reductive emergence.

As a specific example from thermodynamics, picture a liter container of helium gas in thermal equilibrium at room temperature and atmospheric pressure. The helium molecule is a single atom that we can attempt to approximate by a tiny, hard sphere. Imagine a large number of them in a container and assume the molecules act like billiard balls bouncing around inside the container. Using ordinary Newtonian mechanics and statistical averaging we can derive a simple equation called the *ideal gas equation* that relates the pressure, volume, temperature, and number of molecules in the gas. (This is a freshman physics problem in college.) Properties of the whole gas such as pressure and temperature are meaningless for a single molecule. The gas can be thought of as a higher level of complexity with the "emergent principle" stated by the ideal gas equation and "emergent properties" such as pressure and temperature. This works very well in many practical applications.

The principles and properties of an ideal gas follow what is called a "bottom-up" causal arrow. Kauffman also refers to similar situations as having a "top-down" explanatory arrow, where we look to the lower, simpler levels for an explanation of phenomena. This is, of course, the historical procedure of physics going back to Galileo and Newton.

The fact that the ideal gas equation works well over a wide range (but not all) temperatures and pressures shows that the approximation is a good but not perfect one. The equation does not apply to gases out of equilibrium, such as Earth's atmosphere, where we do not have the same temperature and pressure throughout the system. By means of computer simulations meteorologists develop other models, again starting from the basic particle mechanics. Those models also emerge from bottom-up, even if the principles that emerge cannot be mathematically derived from physics.

Another important area of macrophysics is fluid mechanics. Again, its principles are derivable from particle physics. One of the basic equations of fluid mechanics is the Navier-Stokes equation, which describes fluid flow. Kauffman writes that his "physicist friends" tell him that they cannot deduce the Navier-Stokes equation from "more fundamental quantum mechanics" and have "largely given up trying to reason 'upward' from the ultimate physical laws to larger-scale events in the universe."[15]

I am sure he has misunderstood his friends. First of all, the Navier-Stokes equation is a *classical* physics equation and is easily derived from Newtonian mechanics. I have done it in class when I taught undergraduate fluid mechanics. Second, everything in classical mechanics follows as a limit of quantum mechanics and so one can accurately argue that Navier-Stokes is deducible from quantum mechanics.

When we go from thermodynamics and fluid mechanics to the next level of complexity—nonequilibrium systems, such as the atmosphere—we run into the possibility of chaos. Only certain kinds of systems exhibit chaotic behavior. The newly discovered principles of chaotic behavior are also cited as examples of emergence. Take, for example, the fractal behavior of chaotic systems described above. This was not predicted from basic mechanics. Yet it appears in computer simulations that use nothing but basic physics. Once again we have a demonstrable case of reductive emergence. Not everything has to be mathematically derivable from basic physics to be reducible to basic physics.

Summarizing, in the case of reductive emergence we have new principles appearing as systems become more complex. These principles do not apply at the lower level of particle interactions. Yet they are fully reducible to particle mechanics and nothing more. While it is true that we cannot use the traditional mathematical methods of physics to derive most emergent principles, as we did for the ideal gas equation, we might be able to infer them from computer simulations that use nothing but basic physics. Philip Ball has given many wonderful examples in his book *Critical Mass*.[16]

TOP-DOWN

The proponents of emergence are not willing to leave it to reductive emergence. They desperately want to find "something there" besides particles, although for the life of me I don't see what they have against particles. I worked with them all my professional life and found much to like about them, as I hope comes out in my chapters on the subject. Indeed, all of fundamental physics is a wonder to behold and I don't feel a bit diminished as a human being for having emerged from that world of wonder—colored quarks and gluons, strange particles, neutrinos that can travel through Earth and transmute into other types, incredibly precise quantum electrodynamics, Higgs bosons that give matter its mass, spontaneous symmetry breaking that gives matter its organization, and much more.

The doctrine that opposes reductive emergence I defined above as holistic emergence. The basic idea is that the whole is greater than the sum of its parts and that at least some emergent principles, even if resulting from bottom-up causation, have developed the ability to act downward, that is, have the emergent property of top-down causation. Or, as Kauffman describes it, some phenomena at one level have an explanatory arrow that points to a higher rather than a lower level. He rather ponderously gives evolution as an example:

> The explanatory arrows with respect to the emergence of novel functions point upward to the evolutionary emergence of preadaptions via natural selection for nonprestatable functions in nonprestatable selective environments. They do not point downward to string theory.[17]

In other words, natural selection is a high-level emergent principle that explains how lower-level organisms evolve. Those organisms are at a lower level since natural selection is a property of the whole of life. The explanatory arrow is bottom-up, that is (admittedly confusingly), the causal arrow is top-down.

Another writer on emergence, theologian James Haag refers to anthropologist Terrence Deacon, Nobel laureate physicist Ilya Prigogine, and others as arguing that "thermodynamic emergence is not simply a mechanistic story, because the dynamics of molecular interactions are time reversible, while the emergent dynamics are not."[18] This is supposedly then an example of top-down causation.

They are dead wrong on this one, even if one has a Nobel Prize. As we saw in chapter 5, all physical processes are, in principle time, reversible at all levels. No law of physics prevents a punctured tire from spontaneously reinflating. The only reason we do not observe macrotime reversal is purely statistical—it is highly unlikely to happen.

Clayton has stated his preference for holistic emergence (which he calls *strong emergence*) as the process whereby some whole has an active nonadditive causal influence on its parts.[19] Theologian J. Wentzel van Huyssteen has nicely summarized Clayton's notion of holistic emergence and related it to similar views of engineer Stephen J. Kline and chemist Michael Polanyi.[20] Kline cites what he calls *Polanyi's law*: "(1) In many hierarchically structured systems, adjacent levels mutually constrain but do not determine one another; (2) In [some] hierarchically structured systems, the levels of control (usually upper levels) harness the lower levels and cause them to carry out behaviors that the lower levels left to themselves would not do."[21]

I wonder how much of this is trivial. A wheel is made up of particles. Turning the wheel has a top-down causative effect on the particles, moving them in a circle, but nothing very profound is involved. The particles are bound in the wheel, which constrains their motion—just as the walls of a box prevent the particles inside from escaping. Similarly, in the example of the ideal gas discussed above, compressing the gas with a piston raises the average kinetic energy of the particles of the gas. In none of these cases does an action at a higher level produce some new fundamental lower-level principle that cannot be understood in terms of elementary particle interactions.

I have examined considerable writing on the subject of emergence and could not find any specific example worked out in any detail that demonstrated an emergent property that exhibited top-down causality. Computer simulations, at every level, even human social systems, are all based on breaking the system down into parts and programming how those parts interact. The fact that these simulations reproduce many of the emergent properties of higher-level systems makes a case for purely reductive emergence with trivial bottom-up causality, as in the examples above. When the parts of a complex system interact new principles do emerge, but they are not more than the sum of the parts and do not represent any holistic laws of nature that are not already implicit in the laws governing the parts.

A nontrivial case of top-down causality would occur if, hypothetically, the water in a living cell had a different molecular structure than "nonbiological" water. For example, if it could be shown convincingly that homeopathy really works, we might have evidence for just that kind of holistic phenomenon that refutes the reductionist paradigm and proves we are not just particles. Many people would love to see just that happen, but careful scientific studies have repeatedly found no support for such phenomena.

Emergence is just a name for the evolution of complexity out of simplicity, no doubt a notable phenomenon and little doubt that it arises purely from particles of matter.

NOTES

1. John Gribbin, *Deep Simplicity: Bringing Order to Chaos and Complexity* (New York: Random House, 2004), p. xvii.

2. James Gleick, *Chaos: Making a New Science* (New York: Penguin, 1987); Gribbin, *Deep Simplicity.*

3. Hermann Haken, *Synergetics, an Introduction: Nonequilibrium Phase Transitions and Self-Organization in Physics, Chemistry, and Biology*, 3rd rev. enl. ed. (New York: Springer-Verlag, 1983).

4. Philip Ball, *The Self-Made Tapestry: Pattern Formation in Nature* (Oxford: Oxford University Press, 1999).

5. Ilya Prigogine and Isabella Stengers, *Order out of Chaos* (New York: Bantam, 1984); John Briggs and F. David Peat, *Turbulent Mirror: An Illustrated Guide to Chaos Theory and the Science of Wholeness* (New York: Harper and Row, 1989).

6. Philip Ball, *Critical Mass: How One Thing Leads to Another* (New York: Farrar, Straus and Giroux, 2004).

7. William Poundstone, *The Recursive Universe* (New York: Morrow, 1985).

8. Ibid., pp. 197–213.

9. Stephen Wolfram, *A New Kind of Science* (Champaign, IL: Wolfram Media, 2002).

10. Paul Davies, *The Cosmic Blueprint: New Discoveries in Nature's Creative Ability to Order the Universe* (New York: Simon & Schuster, 1988); paperback ed. (Radnor, PA: Templeton Foundation Press, 2004), p. 142.

11. Stuart Kauffman, *At Home in the Universe: The Search for the Laws of Self-Organization and Complexity* (New York: Oxford University Press, 1995).

12. Stuart A. Kauffman, *Reinventing the Sacred: A New View of Science, Reason, and Religion* (New York: Basic Books, 2008). Quotation from jacket.

13. Philip Clayton, *Mind and Emergence: From Quantum to Consciousness* (Oxford: Oxford University Press, 2004).

14. Paul Humphries, "How Properties Emerge," *Philosophy of Science* 64 (1997): 1–17

15. Kauffman, *Reinventing the Sacred*, p. 17.

16. Ball, *Critical Mass.*

17. Kauffman, *Reinventing the Sacred*, p. 141.

18. James Haag, "Between Physicalism and Mentalism: Philip Clayton on Mind and Emergence," *Zygon* 41, no. 3 (September 2006): 633–48.

19. Clayton, *Mind and Emergence*, p. 49.

20. J. Wentzel van Huyssteen, "Emergence and Human Uniqueness: Limiting or Delimiting Evolutionary Examination?" *Zygon* 41, no. 3 (September 2006): 649–64.

21. Stephen J. Kline, *Conceptual Foundations of Multidisciplinary Thinking* (Palo Alto, CA: Stanford University Press, 1995), pp. 115–19; Michael Polanyi, "Life's Irreducible Structure," in *Knowing and Being: Essays by Michael Polanyi*, ed. Marjorie Grene (London: Routledge and Kegan Paul: 1969), pp. 225f.

11

RETURN TO REDUCTION

The reductionist worldview *is* chilling and impersonal. It has to be accepted as it is, not because we like it, but it is because that is the way the world works.

—Steven Weinberg[1]

THE FAILURE OF THE BOOTSTRAP

Philosophers of science have long argued over the criteria that should be applied to determine whether or not some line of thought should be identified as "scientific." While any theory—to be considered scientific—must be rigorously tested against observations, it is always possible that sometime in the future even the most successfully tested theory might be proved wrong. However, once you know that a theory is wrong, you can rule it out. It is falsified. In the 1930s Karl Popper[2] and Rudolf Carnap[3] proposed that *falsifiability* be used as the criterion to distinguish science from nonscience. However, this implies that anything that is falsifiable is thereby science, from astrology to creationism. Like pornography, we know pseudoscience when we see it, and astrology and creationism are pseudoscience.

So, although everything that is falsifiable is not science, any good theory whether or not scientific should be falsifiable. For, as Wolfgang Pauli said when told of a theory that was not falsifiable, "It's not even wrong."

S-matrix theorists never made bootstrap theory actually work. They could not produce any falsifiable predictions that were tested, successfully or unsuccessfully, against experiments. They never proved it right or wrong. But they tried, and so did I.

In 1968 a young Italian postdoctoral fellow named Gabrielle Veneziano, working at the CERN, the Conseil Européen pour la Recherche Nucléaire (European Council for Nuclear Research) laboratory in Geneva, Switzerland, proposed a simple mathematical formula that had many of the features desired for the S-matrix. The formula was called the *beta function* by mathematicians, and had been studied by the eighteenth-century Swiss mathematician Leonhard Euler.[4]

In 1968 I was enjoying my first sabbatical at the Institute für Hochenergiephysik (Institute for High-Energy Physics) of the University of Heidelberg in Germany. A good part of that year (while not sightseeing from one end of Europe to the other) was spent trying to test Veneziano's model against the data on the scattering of negative kaons from protons over a wide range of energy. This reaction was characterized by a number of so-called resonances that appeared as peaks in the scattering probability and were interpreted as new, short-lived particles. The hope was that these resonances would show up as "poles" in the beta function. Unfortunately, I was not able to get the model to fit the data very well. The attempts by others to follow a similar procedure were equally unsuccessful.

In science, research that leads to negative results is usually still of some value. The years of effort made by dozens of physicists on S-matrix theory were not a total waste. Physicists discovered that the beta function of the Veneziano model described elementary objects that were not three-dimensional points but were extended into another curled-up dimension.[5] These objects were tiny strings. So, while S-matrix theory died on the vine, the Veneziano model turned out to be the precursor of *string theory*, the candidate for a *theory of everything* (TOE) that has occupied most of the theoretical physics world for the last two decades. So far that quest has been unsuccessful and some are beginning to question whether this is the path that we should be following to seek ultimate unification.

QUARKS

In chapter 9 I told of the particle explosion that occurred in the 1960s as accelerators moved to higher and higher energies. Most of those new particles were very short-lived, strongly interacting particles dubbed *hadrons.* Physicist Murray Gell-Mann discovered that the hadrons fit into certain symmetry patterns he originally called "The Eightfold Way," after the Buddha's path to enlightenment. In *The Tao of Physics,* Fritjof Capra likened these symmetry patterns to those found in Eastern symbols such as the Chinese yin-yang diagram and the Tibetan Buddhist ephemeral sand painting called a mandala.[6]

Gell-Mann found that he could explain the way hadrons were organized by postulating that they were composed of *fractionally charged* particles, which he whimsically named *quarks.* Later he found the word in James Joyce's *Finnegans Wake,* "Three quarks for Muster Mark!" The quark idea was also arrived at independently by George Zweig at about the same time.

At first Gell-Mann did not insist quarks "really existed," but suggested they were simply calculational tools. However, evidence for pointlike, fractionally changed structures inside protons and neutrons was found in experiments conducted at the Stanford Linear Accelerator (SLAC) in Palo Alto, California, during the period 1966–78. In these experiments, very high-energy electrons were able to probe inside the protons and neutrons in a manner similar to the way Ernest Rutherford probed the gold atom with alpha particles in 1909. In both cases the probing particles scattered at large angles, signaling the existence of tiny scattering centers inside the object being probed. Later the SLAC results would be confirmed at the Fermi National Accelerator Laboratory (Fermilab) in Batavia, Illinois, using neutrinos as probes. I collaborated in several of these experiments as a member of the University of Hawaii particle research group.

The reason probing the structure of protons and neutrons worked with electrons or neutrinos as probes, and did not when protons or other hadrons were used, is that electrons and neutrinos are pointlike at these energies. To see pointlike objects you need pointlike probes. Hadrons, including the proton and the neutron, were broken down into smaller parts—quarks. And no one knew how to produce a beam of free quarks.

THE STANDARD MODEL

It was found that two quarks, the *u*-quark and *d*-quark ("up" and "down" in current terminology), combined in groups of three to make up the proton and the neutron. The proton is *uud*; the neutron is *udd.* For each quark there corresponds an *antiquark* of opposite charge, $\bar{u}$ and $\bar{d}$. The antiproton and the antineutron are made of these antiquarks. Other particles, called *mesons*, are made of quark-antiquark pairs. The negative K-meson, or *negative kaon*, is $K^- = \bar{u}s$, where *s* is a *strange* quark. The positive kaon I studied for my PhD is $K^+ = u\bar{s}$, where $\bar{s}$ is an *anti-strange* quark. It was found that all of the hundred or so hadrons discovered in the sixties appeared to be composed of the *u*, *d*, and *s* quarks, their antiquarks, plus three additional heavier quarks and antiquarks: *c* ("charm"), *t* ("top," or "truth"), and *b* ("bottom," or "beauty").

Besides the hadrons, the electron is accompanied by two heavier versions, the *muon* and the *tauon*—each with one unit of negative electric charge. Each are accompanied by a zero-charge neutrino. Together they form another class of particles called *leptons* that are themselves elementary in the standard model. Each lepton also has an associated antiparticle.

The standard model picture is summarized in table 11.1, where we see

	Fermions (antiparticles not shown)			Bosons
Quarks	u	c	t	γ
	d	s	b	g
Leptons	ν_e	ν_μ	ν_τ	Z
	e	μ	τ	W

Table 11.1. The elementary particles of the standard model. Shown are the three generations of spin 1/2 fermions—quarks and leptons—that constitute normal matter and the spin 1 bosons that act as force carriers. Each has an antiparticle, not shown. The electroweak force is carried by four spin 1 bosons—γ, W^+, W^-, Z, and the strong force by eight spin 1 *gluons*, g. A yet-unobserved spin zero *Higgs boson*, not shown, is also included in the standard model.

three "generations" of *fermions*, each containing two quarks, two leptons, and their antiparticles. The top row of quarks has +2/3 the unit electric charge. The second row of quarks has charge –1/3. Their antiquarks have opposite charge. The top row of leptons is composed of three types of neutrinos, which are electrically neutral. The second row of leptons has one unit of negative electric charge. The antileptons have the opposite, one unit of positive electric charge.

Each succeeding generation is heavier than the preceding one, and so the heavier quarks and charged leptons tend to decay into the lighter ones and so are not seen in normal matter.

Fermions are particles with half-integer intrinsic angular momentum, or *spin*. Particles with integer or zero spin are called *bosons*. The three generations of fermions and their antiparticles in the standard model have spin 1/2. The four bosons, which are given in the right column, have spin 1.

The bosons are often called "force particles" because they mediate the fundamental interactions of quarks and leptons, acting as "messengers" between quarks and leptons at different points in space. These include the photon, the particle of light, which is responsible for the electromagnetic force (indicated by the Greek symbol γ); three *weak bosons* that account for the weak nuclear force (W^+, W^-, Z); and eight *gluons*. In the standard model, electromagnetism and the weak nuclear force are united into a single *electroweak* force. One additional spin zero particle called the *Higgs* boson is predicted by the standard model but has not yet been produced in the laboratory. If it remains unseen in the new generation of colliding beam accelerators now coming into operation, the standard model will be falsified, or will at least need to be drastically revised.

THE BASIC FORCES

As discussed in chapter 9, in the late 1940s the quantum theory of electromagnetism known as quantum electrodynamics (QED) was developed. QED describes the interactions of charged particles, such as electrons and quarks, with photons. The basic interaction between electrons, for example, is understood as the exchange of a photon between the two charged particles, as was illustrated in figure 9.1 (a). The photon carries energy and momentum from one particle to the other. More-complicated exchanges, such as those shown in figure 9.1 (b) and (c), also must be taken into account.

Following the success of QED, the weak nuclear force was described in terms of the exchange of three additional "weak" bosons: W^+, W^-, and Z, with electric charges +1, –1, and 0, respectively, in terms of the unit electric charge (see figure 9.2). Unlike the photon, which has zero mass, the weak bosons are very massive. The Ws have 85 times the mass of a proton, and the Z has 97 times the proton's mass.

In the 1970s electromagnetism was combined with the weak nuclear force into a single unified entity called the *electroweak* force. The unification scheme, attributed mainly to Steven Weinberg, Sheldon Glashow, and Abdus Salam, worked brilliantly, predicting the masses of the weak bosons.

About the same time a theory of the strong nuclear force called *quantum chromodynamics* (QCD) was developed that successfully described the interactions of quarks, although it has never been tested to high precision. QCD was also patterned after QED, the strong force between quarks being described as the exchange of massless particles called *gluons* (they are the "glue" that holds the nucleus together). Eight gluons are needed.

In addition to fractional electric charge ($\pm 1/3$ or $\pm 2/3$ of the unit electric charge), quarks carry a quantity called *color charge* that, in analogy to the familiar primary colors, are called *red*, *green*, and *blue*. Anti-quarks carry the complementary color charges, *cyan*, *magenta*, and *yellow*. All of the hadrons that appear in nature are "white" in terms of color charge. Those that are formed from three quarks, such as the proton and the neutron, are *red* + *green* + *blue* = *white*, while the mesons formed from a quark and an anti-quark are *red* + *cyan* = *white* and similar combinations. Since they are not *white*, free quarks are not found as free particles in nature.

Figure 11.1 illustrates the basic interactions in the standard model. In (a) we have the electromagnetic interaction between an electron and a quark via photon exchange. Similar diagrams can be drawn for the electromagnetic interactions of any of the charged particles in table 11.1—quarks interacting with quarks, electrons with muons, and so on.

In (b) the weak interaction is illustrated in terms of the decay of a *d* quark into a *u* quark, electron, and anti-electron-neutrino. This is the basic *beta-decay process*, where the quarks are inside the nuclei. In (c) the strong nuclear force between two quarks via gluon exchange is illustrated.

Note that all these diagrams are the simplest we can draw. They are good approximations in the cases of (a) and (b). However, to accurately describe observations between hadrons, which contain quarks, more diagrams than the one shown in (c) are required, with gluons being emitted and absorbed by various quarks.

But that is getting us into far more technicalities than are needed for this book. I have gone into the detail that I have to assure the reader that the reductionist picture is now once again firmly established in physics. The standard model just described has agreed with all the data taken in

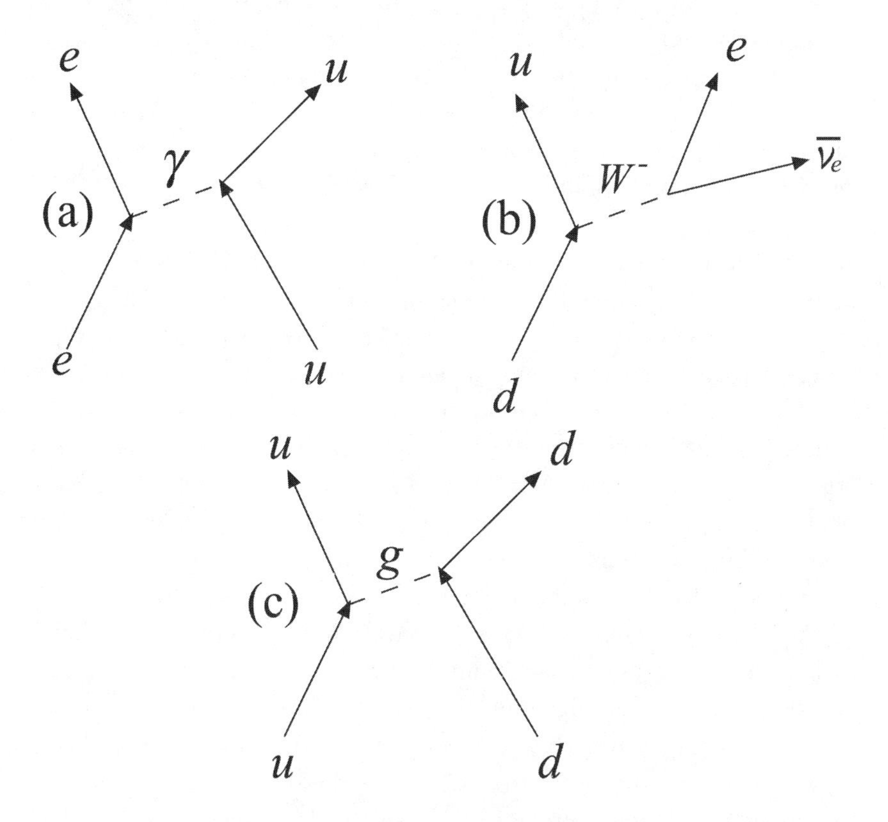

Fig. 11.1. Examples of the three basic forces between particles in the standard model, illustrated with Feynman diagrams. In (a), the electromagnetic force between an electron and a *u*-quark is mediated by the exchange of a photon, indicated by the symbol γ. In (b) an example of the weak nuclear force is shown with the *beta-decay* of a *d-quark* into a *u*-quark, electron, and antineutrino is mediated by the weak force boson W^-. In (c) the strong nuclear force between a *u*- and a *d*- quark is shown as mediated by a *gluon*.

physics laboratories for more than thirty years. Furthermore, it has also given us great insight into the early universe, where the same physics is known to apply back to when the universe was only 10 millionths of a second old.

Suffice it to say in summary that normal matter, as we know it, is constituted from a total of six types of quarks, six antiquarks, six leptons, six antileptons, twelve force bosons, and a Higgs boson. All these particles, with the exception of the Higgs, have been observed either directly or indirectly and their masses and other properties established. The Higgs boson remains to be either observed in the new collider experiments or shown to not exist. The latter outcome would not be too disappointing to physicists since it would point the way to new physics beyond the standard model in a way that simple confirmation would not. We will have to wait and see, but thankfully not too much longer.

Ironically, physics has not had much to cheer about now for a quarter century because the standard model has worked so well, without a single empirical anomaly disagreeing with the model in all that time. This is unprecedented in physics, where there had always been at least a few mysteries to whet our appetites and encourage speculation.

The energies required to produce most of these particles do not occur naturally in the current "cold" universe. Some will appear briefly when high-energy cosmic rays from outer space hit the top of Earth's atmosphere. As mentioned above, few of the products of these collisions reach Earth except for muons and neutrinos. Most have been observed in the controlled experiments done with particle accelerators.

So, for most practical purposes in everyday life, three types of elementary particles suffice to make up the matter involved. These are the *u*- and the *d*-quarks that constitute the nuclei of atoms around which the atom's electrons congregate. In addition we have the photons that produce light and provide the attractive force that holds atoms and molecules (groups of atoms) together and the gluons that hold nuclei together. The weak bosons play a small role on Earth, accounting for a portion of the natural radioactivity by way of the beta-decay of nuclei. However, they provide our main source of energy by supplying the mechanism for the nuclear reactions that take place at the center of the sun.

THE DEATH OF HOLISM

Recall that in *The Tao of Physics*, Capra placed great emphasis on what he claimed was the interconnectedness of the universe and how it exists as one indivisible whole. As he put it,

> Quantum theory thus reveals a basic oneness of the universe. It shows that we cannot decompose the world into independent units.[7]

Capra was putting his money on bootstrap theory, which did away with the notion of elementary particles. However, even by the time of the publication of *Tao*, elementary particles were back in fashion and had become part of the standard model.

As the holistic S-matrix waved good-bye, physics reverted back to the old reductionism that had always served it so well. Capra tried bravely to find a parallel between quark symmetries and Eastern philosophy, but it was clear that the teachings of Hinduism, Buddhism, and Taoism had no more to do with modern physics than did the Torah, the New Testament, and the Qu'ran. The stories in sacred books do not bear even the most superficial resemblance to modern physics and cosmology. They present a deeply contradictory picture, despite the strained efforts of theologians and theistic scientists to claim some remote connection.[8]

Capra's contention that physics implies a universe as one interconnected whole is edifying and poetic. Certainly, various parts of the universe are connected in various ways. But Capra insists that we cannot understand our world unless we treat it as one inseparable whole.

Apparently that's not the way things work. First of all, trying to understand everything all at once is probably impossible. Second, it is not necessary. The standard model has shown once again that an effective way to formulate a useful model of the universe is by taking it apart, piece by piece, and then studying these pieces by themselves under controlled conditions.

This is true also for quantum theory, which Capra also clearly had in mind in formulating his views. In the early twentieth century, scientists did not attempt to use the new quantum theory to calculate the energy levels of an atom of each element in the Periodic Table. They started with the simplest atom, hydrogen. And when they had a good understanding of that, they moved to the next simplest—helium. The properties of heavier (and hence more complicated) atoms proved increasingly difficult to calculate, having more and more electrons that interacted with one another and with the nucleus in complex ways. So atomic physicists did the best they could, with a combination of theory and experiment, applying the general principles they had learned in the simpler cases to the more complicated ones. The result was a good understanding of atoms that then could be applied to the even more complex structures—molecules, that are formed from atoms, especially the long chains of proteins and nucleic acids that constitute life.

Note that the atomic theory of matter and the quantum theories of both matter and light introduced discreteness or lumpiness where previously we had continuity or smoothness. Quantum spiritualists will often claim that quantum mechanics restored "continuity" to physics, replacing reductionist atomism with a new holism. In fact the opposite is the case. We now break the universe down into far more parts than ever dreamed of in the nineteenth century.

Taking a holistic view of the human body, as ostensibly is done in Eastern medical practices, may be emotionally soothing. But in practical terms it gains us little benefit, as evidenced by the inability of unscientific Chinese or Hindu medicine to come close to the success of Western, science-based medicine in the curing of illness. With science-based medicine, a part of the body that is malfunctioning is treated reductively, with some lip service provided by the patient's health maintenance organization that they "treat the whole person"—since that is what people apparently want to hear.

In the meantime, Fritjof Capra continued on his holistic crusade and was considered a major guru for what came to be called the *New Age* of human spirituality. His other books and a film have focused on humanity working together to achieve global harmony both socially and environmentally. At this writing he is the founding director of the Center for Ecoliteracy in Berkeley.

SO WHAT'S THE TAO?

Despite the success of the standard model, no one thinks of it as the final word—the *tao*—or principle behind everything. The next generation of high-energy particle colliders at CERN and Fermilab is expected to produce the first deviations from the standard model to be seen in a generation. Physicists will be very unhappy if it does not. The absence of empirical anomalies has crippled the efforts of theorists to develop a theory that moves to the next level beyond the standard model.

In chapter 4 we saw how an early attempt in the 1970s at a *grand unified theory* (GUT) was picked up by Maharishi Mahesh Yogi as a basis for his teaching about cosmic consciousness. He mirrored Capra's use of high-energy physics to provide a connection with Eastern mysticism. This claim met the same fate of being falsified by the data from the physics laboratory.

NOTES

1. Steven Weinberg, *Dreams of a Final Theory: The Search for the Fundamental Laws of Nature* (New York: Random House, 1992), p. 53.

2. Karl Popper, *The Logic of Scientific Discovery*, English ed. (London: Hutchinson; New York: Basic Books, 1959). Originally published in German (Vienna: Springer Verlag, 1934).

3. Rudolf Carnap, "Testability and Meaning," *Philosophy of Science* B3 (1936): 19–21, B4 (1937): 1–40.

4. For the mathematically savvy, the beta function is simply the product of two gamma functions divided by another gamma function.

5. Leonard Susskind, *The Cosmic Landscape: String Theory and the Illusion of Intelligent Design* (New York: Little, Brown, 2006); Lee Smolin, *The Trouble with Physics: The Rise of String Theory, the Fall of a Science, and What Comes Next* (New York: Houghton Mifflin, 2006).

6. Fritjof Capra, *The Tao of Physics* (Boulder, CO: Shambhala, 1975), pp. 255–61.

7. Ibid., p. 68.

8. Gerald L. Schroeder, *The Science of God: The Convergence of Scientific and Biblical Wisdom* (New York: Free Press, 1997); see my review in Victor J. Stenger, "Fitting the Bible to the Data," *Skeptical Inquirer* 23, no. 4 (1999): 67–68; Frank J. Tipler, *The Physics of Christianity* (New York: Doubleday, 2007); see my review in Victor J. Stenger, "In the Name of the Omega Point Singularity," *Free Inquiry* 27, no. 5 (August/September 2007): 62; Taner Edis, *An Illusion of Harmony: Science and Religion in Islam* (Amherst, NY: Prometheus Books, 2007).

12

GHOSTBUSTING THE QUANTUM

It is impossible for anyone to dispel his fear over the most important matters, if he does not know what is the nature of the universe but instead suspects something that happens in myth. Therefore, it is impossible to obtain unmitigated pleasure without natural science.

—Epicurus

WHAT DO THE DATA SAY?

In chapter 8 we summarized the history of quantum mechanics and discussed its main principles. We saw why many people, including Einstein, have labeled it "spooky." Quantum mechanics involves phenomena that defy common sense and threaten to overthrow the traditional reductionist methodology of sciences, ranging from physics to medicine in which the subject matter is broken down into parts that can be treated independently. Even more revolutionary, quantum mechanics seems to involve a special place for the human mind in controlling the very nature of reality and doing this instantaneously over the entire universe and back in time as far as time is measured.

In this chapter and the next, we will examine what the conventional theory of quantum mechanics actually does say about these issues. We will see how the themes of quantum spirituality have come about at least partly from a gross misinterpretation of the nontechnical language used in trying to provide quantum mechanics with some philosophical foundation, and in trying to explain it to laypeople. Words are always subject to dispute as to their meanings, and this has led to honest confusion at best or purposeful misrepresentation for financial gain at worst.

In this chapter we will also take a look at the data to see if there is an empirical basis to the notion that the human mind has some spiritual element that enables it to go beyond the world of matter.

As discussed in chapter 2, physicist Amit Goswami was singled out among those contributing to the film *What the Bleep Do We Know!?* as the physicist whose views are the most antithetical to both Western science and common sense. But he is still a physicist and recognizes the importance of observational testing in science. He claims that evidence for special powers of the mind can be found in psychic studies such as *mental telepathy* (*extrasensory perception*, or ESP) and *mind-over-matter*. He says, "Experiments [proving telepathy] have been carried out in many different laboratories and positive results are claimed."[1] Goswami specifically refers to work done by Harold Puthoff and Russell Targ,[2] and Robert Jahn.[3]

However, Goswami is forced to acknowledge, "Telepathy has not yet been recognized as a scientific discovery." Nor has it been today, I might add, a quarter century after the references he cites. Goswami gives as a reason for the skepticism about ESP: "It does not seem to involve any local signals to our sense organs and hence is forbidden by material realism."[4] While this certainly is a reason for skepticism, it has not retarded psychic studies. Even if signals are transmitted by some unphysical mechanism, the effects of those signals should be observable by our material eyes and instruments.

Indeed, psychic researchers have long claimed, as Goswami says above, that they have scientific evidence for psychic phenomena. To be honest, they must admit that such evidence has not been accepted by the consensus of the scientific community. I have discussed psychic studies in several books and concluded from the lack of convincing evidence over a long period of time and from the total absence of statistically significant replicability that the phenomena can be declared nonexistent beyond a reasonable doubt.[5] I will summarize the basis of this conclusion in the next section.

PSYCHIC SCIENCE

If some special vital force, spirit, or soul exists, it should manifest itself in mental phenomena that go beyond what we associate with the material world and describe by physical law. Examples of psychic phenomena include extrasensory perception (ESP); mind-over-matter, or psychokinesis (PK); out-of-body experiences (OBEs); and near-death experiences (NDEs), which are actual returns from the dead. All have been extensively studied in the laboratory, in some cases for more than a century.

In 1853 one of the greatest scientists of all time, Michael Faraday (d. 1867), set up a test for "table turning." This is the séance experience in which people sit around a round table with their palms flat down on the tabletop. Despite the fact that they make no conscious effort to move the table, it rotates under them. This was regarded at the time as a kind of mind-over-matter, with the combined "psychic energy" of the participants turning the table.

Faraday prepared his experiment by attaching several layers of thick paper to the tabletop, using a soft cement that allowed some movement between layers. He found the top layer rotated more than the layer cemented to the tabletop, proving that it was the participants' hands and not the table doing the moving.[6] This unconscious act of hand movement is a natural explanation for the familiar Ouija board game, trademarked by Parker Brothers, where the *planchette* acts as a movable indicator to answer questions.

Faraday's debunking of table turning hardly put an end to the scientific investigation of psychic phenomena. In fact, it marked the beginning of a field of investigation that has continued to the present day.

While no major religion makes claims that humans possess ESP, if we go back through the early history of psychic studies we see signs that most investigators were motivated by personal Christian religiosity to find evidence for the soul or an immaterial force of some kind. After all, wireless telegraphy had been demonstrated; why not wireless telepathy? Certainly religion was the motivation for the two most prominent psychic investigators of the late nineteenth century, William Crookes and Oliver Lodge.[7]

William Crookes discovered the chemical elements thallium and selenium and invented the vacuum tube used by J. J. Thomson to discover the electron. After his brother's death in 1867, Crookes began an attempt to scientifically demonstrate the reality of the powers of the spiritualist mediums who were much the rage in England and America at the time. Most had started out as stage magicians but found it more lucrative to hold séances in which they would demonstrate their supernatural powers (always in the dark, of course) and enable gullible patrons to speak to dead loved ones. Curiously, the dead never provided the listeners with any useful information about the future, such as a winning horse or lottery number, which they surely would have had access to in timeless heaven.

Ignoring what he must have known were the proper protocols of good science, Crookes performed poorly controlled experiments with demonstrably fraudulent mediums that he felt confirmed his belief in a spirit world. One fatal mistake was allowing his subjects to control the experiments. Several of his subjects were caught cheating by others and skeptics effectively debunked all his claims.[8]

Oliver Lodge transmitted radio signals before Marconi (but after Tesla), and he invented spark-plug ignition as well as the vacuum tube used in electronic circuits and other devices that were utilized well into the mid-twentieth century.

Like Crookes, Lodge was another highly competent scientist who allowed his personal belief in spirits to override any natural skepticism he would be expected to exhibit given his high stature in the field. His son Raymond had been killed in Flanders in 1915 and Lodge insisted that Raymond was communicating with his family from the world beyond. Lodge also allowed himself to be bamboozled by phony psychics.[9]

The third major figure in the history of psychic research was Joseph Banks Rhine of Duke University, who in the 1930s made an honest attempt to do careful laboratory tests of paranormal phenomena. Although provable cheating by research assistants occurred in his lab (Rhine tended to keep those who seemed to have psychic powers and dismiss those who didn't), Rhine was never accused of personally faking data. Rather, like Crookes and Lodge, he allowed his personal beliefs to cloud his better judgment. Rhine was a religious man and hoped his work would help reconcile science and religion by finding scientific evidence for a nonphysical component in humans. He clearly did not regard spiritual phenomena to be beyond the capabilities of science to observe.

Rhine coined the term "ESP" and founded the *Journal of Parapsychology* after his submissions to conventional journals continued to be turned down. Papers in the new journal were peer reviewed, but by ESP sympathizers. Again, Rhine's claims for positive evidence did not hold up under critical scrutiny from sources outside his little club of supporters.[10]

In more recent years we have seen a group at Princeton led by engineer Robert G. Jahn claim statistically significant results for mind-over-matter, including the ability to act back in time. None were ever independently replicated.[11] This effort just lately ceased operations.[12]

I have given just a small sample of the psychic investigations that have gone on for a century and a half. In all that time, psychic investigators, also known as parapsychologists, have never come up with positive results that stood up to the level of critical analysis applied to all scientific claims. In every other field that I can think of, such a sustained record of negative results over so many years would have long ago resulted in the sought-after phenomenon being declared nonexistent. Goswami has no basis for inferring superpowers of the mind from quantum mechanics or claiming that they are empirically verified. In fact, the data now show beyond any reasonable doubt that these powers do not exist.

PSI AND THE SUPERNATURAL

Today's parapsychologists are usually very careful to deny that they are drawing any supernatural conclusions whenever they claim evidence for a paranormal effect. But let us imagine what would happen if well-controlled experiments on ESP should find evidence that passes the most stringent tests that science can provide, leading even the most skeptical to admit that the phenomenon is real. You can rest assured that parapsychologists and theologians would readily change their tune about whether established evidence for ESP was evidence for a human soul. How much more important to have discovered evidence for the soul—for a world beyond matter—than just another form of physical communication! Even nonbelieving scientists would have to accept what the data are telling them. Most would be happy to do so, with all those opportunities for well-funded research opening up to them. Scientists of every stripe would initially seek to explain the results in terms of known natural processes and perform experiments that would test that hypothesis.

For example, the strength of the signal would be measured as a function of all kinds of variables such as direction and distance from the source.

If it fell off with the square of the distance, then this would be evidence that ESP emissions are a form of conserved energy and probably a natural phenomenon.

Indeed, at the suggestion of Einstein, Rhine performed such an experiment. When he found no "distance effect" he concluded that the phenomenon was not physical. Although the signal was insignificant at every point, Rhine avoided the more obvious conclusion that the phenomenon was not real.

If that experiment were duplicated with a clear ESP signal that did not fall off with distance, then we would have evidence that ESP violates energy conservation and does not behave as expected for a material force. While it would undoubtedly be argued that this result could still be natural, energy conservation is such a fundamental law of matter that its violation would be strong evidence for a reality beyond matter.

But, the facts are that no significant evidence of ESP or other psychic phenomena has ever been found. This statement will be disputed by many workers in the field who argue that significant results have been reported, most recently by Dean Radin.[13] Here are three of the many reasons why the rest of the scientific community has been unwilling to accept these claims:

1. The reported statistical significances for the signals are not adequate for an extraordinary claim. For example, a positive effect in parapsychology is typically claimed when the statistical significance is one in twenty. That is, if the same experiment were to be repeated many times, one in every twenty would produce the same signal or a greater one as the result of a statistical fluctuation. This means that every twentieth paper that is published is a statistical artifact; probably even more since negative results are often just stuffed in a file drawer. In my field of particle physics, the best journals will not publish any extraordinary claim unless the statistical significance is one in ten thousand.
2. The claim is made by Radin and others that a *metanalysis* of several experiments that are not statistically significant individually, those in the file drawer, yields significant results. Metanalysis is notoriously unreliable.[14] I do not know of a single example of a new phenomenon, extraordinary or ordinary, that was first discovered by metanalysis.

3. In no case has an experiment claiming a positive signal for a psychic phenomenon been replicated. While replications are claimed, they are never with the same or closely similar experimental setups and at the same quantitative level.

THE HUMAN ENERGY FIELD

The notion that the human body contains some special energy field called *qi* or *ch'i* in China, *ki* in Japan, and *prana* in India, that contains the force of life is a common teaching in Eastern philosophy and is a major part of the new spirituality. This provides a common thread that runs through most of *complementary* or *alternative medicine* (CAM). This vital force is easily connected with the Western concept of soul.

Millions of dollars are spent yearly on chiropractic, homeopathy, acupuncture, therapeutic touch, and traditional healing practices from China and India. None of these healing methods has a scientific basis in either theory or valid empirical data, being based mainly on anecdotes and prescientific superstitions. Some, like homeopathy, violate well-established principles such as the atomic model of matter. After all the dilutions gone through in the preparation of a homeopathic remedy, the chances are negligible that a single atom of the active ingredient remains.

Recent speculations in the CAM literature have concentrated on finding a place for the vital force within existing science. While different designations are used, a common descriptive term that I will adopt is *bioenergetic field*. This emphasizes first of all that the force is uniquely biological. Second, the force is some kind of energy. Third, the vital force is a field, that is, it is not localized in space but spreads out "holistically" from its source.

Now, each of these three properties provides a means for testing the concept. If the force is purely biological, then it should only be found in living things and not be present in inanimate objects or dead organisms. If the force is a form of energy, it should exhibit energy-like behavior, such as it being conserved. If the force is a field, then we should be able to map it like the electric field around a point charge.

What are the facts? Biologists have never identified any components in living organisms that are not composed of the natural elements and no force other than those of the standard model of particles and forces. As for the other testable properties, no laboratory experiments have yet even isolated a bioenergetic field, much less mapped it out and determined if it is conserved.

One way to test if bioenergy is conserved is to see if it falls off with distance from the source. We saw above that Einstein had suggested this as a test for ESP and that experiments by Rhine failed to show any "distance effect." The only data I know of on *qi* I reported on in *God: The Failed Hypothesis* and presented at several universities in China in 2005. As with ESP, the claimed effect did not fall off with distance.[15]

Some of the CAM literature associates the bioenergetic field with electromagnetism. In that case, it is not unique because electromagnetism is not limited to living matter. Now, as we saw in chapter 8, every object emits blackbody radiation—electromagnetic radiation whose wavelength spectrum depends on the temperature of the body. Figure 8.4 (p. 115) showed the spectrum for a human being at normal body temperature. A common scam at psychic fairs is to take your picture with infrared film so that you can see your surrounding "aura." It is no different from the aura of a rock at the same temperature. In any case, since it is simply composed of purely random thermal radiation and not unique to living things, our blackbody radiation is hardly a candidate for the bioenergetic field.

The human body does contain oscillating electrical currents that will generate electromagnetic fields. These can come from the heart, seen in an electrocardiogram, or from the brain, seen in an electroencephalogram. But these fields are very weak and require detectors placed on the skin. They have no effect at any distance away. Correspondingly, someone such as a touch therapist could hardly have any ability to "manipulate" this field from a few inches away. Sadly, therapeutic touch is widely taught in nursing colleges in the United States today without the slightest scientific basis.[16]

PHOTONS ARE NOT WAVES

Let us now take another look at quantum mechanics to see if the theory itself really implies the spooky interpretations given to it by the quantum spiritualists. We begin with the so-called *wave-particle duality*, which is interpreted by quantum spiritualists and some physicists to mean that whether an object is a particle or a wave depends on what the observer, presumably a conscious human being, decides to measure. This leads to the dictum that we have seen as part of New Age spirituality since the 1960s—that we make our own reality.

In chapter 8 I discussed this question in terms of the double-slit experiment, which was first used by Thomas Young in 1800 to demonstrate the wave nature of light. As we have seen, interference is also observed for particles such as electrons, which implies that particles also have wavelike behavior. Furthermore, light itself is now known to be particulate in nature, occurring in localized bits of matter called photons.

Let us set up a double-slit experiment with equipment not available to Young, or indeed to early twentieth-century investigators. Let the surface illuminated by the light from the slits consist of an array of photon detectors sensitive at the one-photon level, as shown in figure 12.1.

As seen in figure 12.2(a), we get individual, localized hits just as

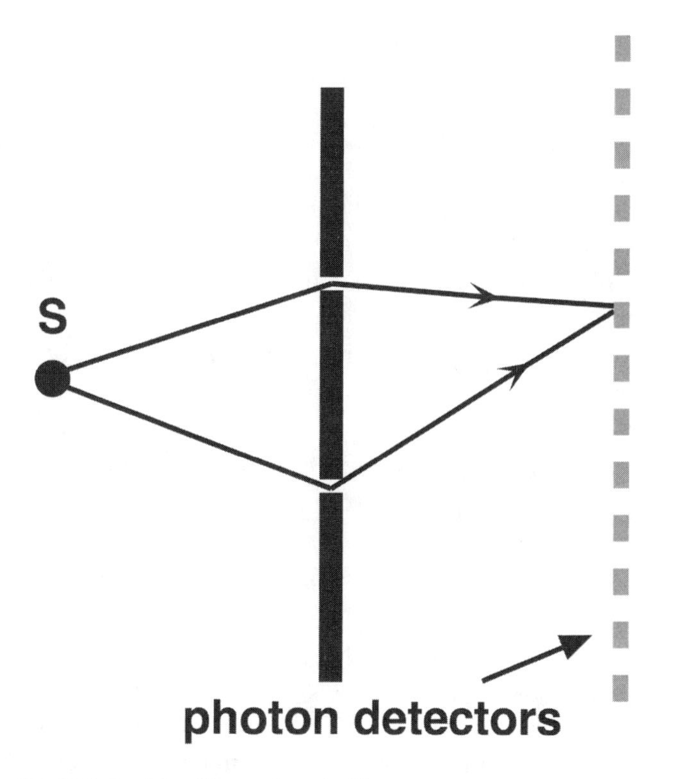

Fig. 12.1. Double-slit experiment with detectors able to register single photons.

expected for particles. When the number of photons is small, the hits look more or less random. But as you accumulate data in (b) and (c), a fascinating thing happens. The pattern of hits takes the shape of the interference pattern expected for waves. But there are no waves, just particles. The same experiment can be done with electrons or other particles, showing the same effect. This allows us to conclude that the basic objects, whether contained in a light beam or a beam of electrons, protons, or any other body in nature, are particles. There is no wave-particle duality. Photons are just particles.

As Richard Feynman put it in a lecture to high school students:

> I want to emphasize that light comes in this form—particles. It is very important to know that light behaves like particles, especially for those of you who have gone to school, where you were probably told something about light behaving like waves. I'm telling you the way it does behave—like particles.[17]

THE FICTIONAL WAVE FUNCTION

So, then, what is waving? What is the source of the observed interference pattern that fits what is expected for waves? That pattern is the statistical distribution of a large ensemble of individual particle detections. In the Schrödinger wave formalism of quantum mechanics, going back to 1926, the wave is represented by an abstract mathematical quantity called the *wave function.* That wave function is normally taken to be a *complex function.* At a given point in space and moment in time this function has a value given by a complex number ψ. The simplest way to describe a complex number such as ψ is as a two-dimensional vector, as illustrated in figure 12.3. The length of the vector is the *magnitude* of ψ and the angle with respect to one axis is called the *phase of* ψ. The projections of the vector on the two axes are called the *real* and *imaginary* parts of ψ.[18] In quantum mechanics the square of the magnitude of ψ gives the probability for finding a particle at a certain point in space at a certain time, per unit volume.

If you insist on interpreting the wave function as a "real" physical entity such as a water wave, then it moves faster than the speed of light, indeed, infinite speed, violating a basic tenet of Einstein's special theory of

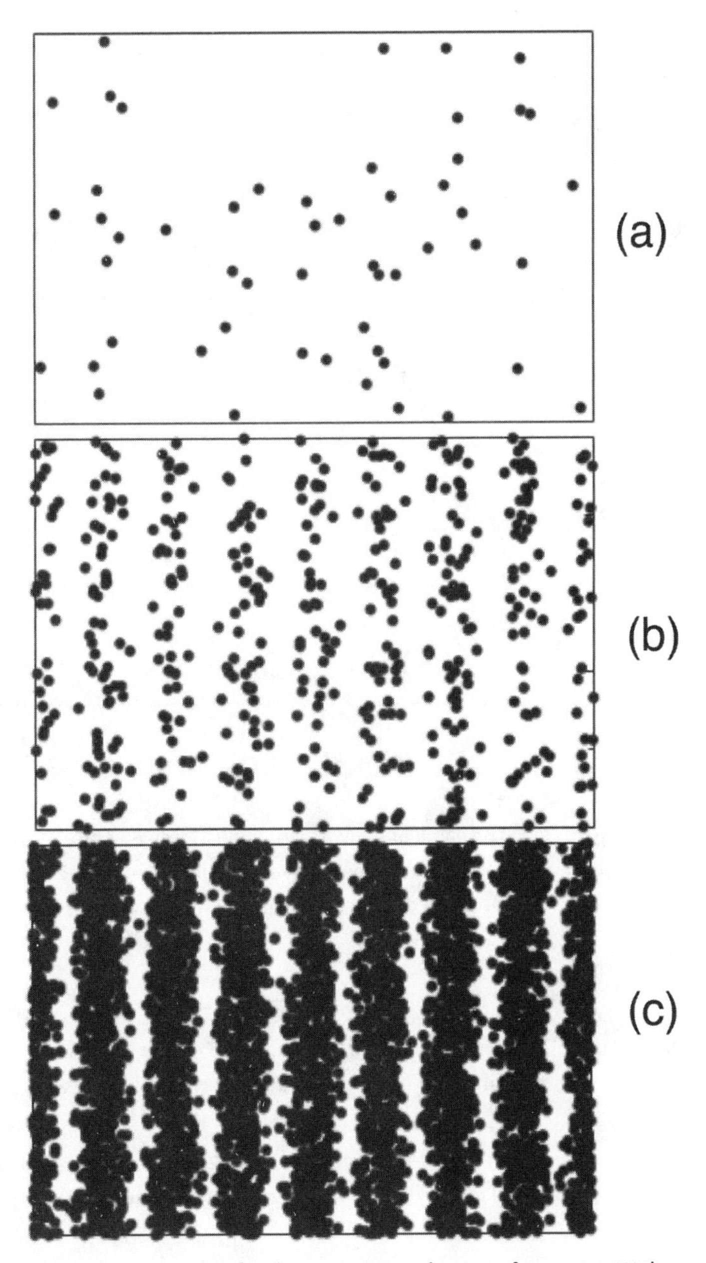

Fig. 12.2. Photon hit pattern for the experiment shown in figure 12.1. With just a few hits (a) we see what looks like a random localized particle hit pattern. But as the number of hits increases in (b) and (c), the familiar double-slit wave interference pattern emerges as a statistical effect.

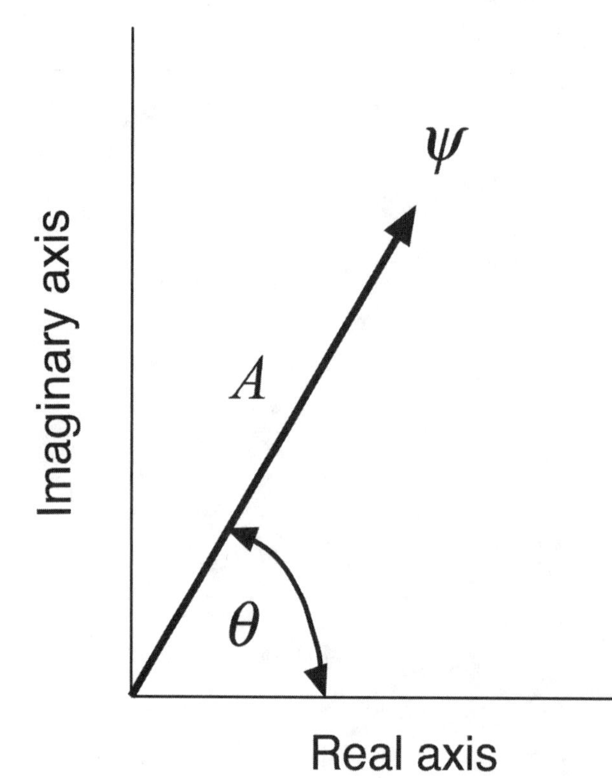

Fig. 12.3. The value of the wave function at a given point in space and moment in time is represented by a complex number ψ. It can be described as a two-dimensional vector with a *magnitude A* and a *phase θ*.

relativity. However, if we accept that the wave function is just an abstract mathematical entity physicists use to compute the probability for finding a particle at a particular position in space, then there is nothing spooky about it. Abstract things can move as fast as their inventors wish.

Actually, the term *wave function*, though still widely used, is a misnomer. It does not describe a vibration of any kind of medium and need not even be mathematically a wave. A much better designation is *state vector*. This is a vector in an abstract space that specifies the state of a system. A simple example of a two-dimensional abstract space is given in figure 12.3.

Consider the following example, which is illustrated in figure 12.4.

Suppose you are a resident of a planet in the Alpha Centauri multiple star system, the system nearest to our solar system. Back on Earth, a friend enters your name in a lottery where the prize is a million dollars and your chance of winning is one in a million. If you win the lottery, your probability of winning changes instantaneously to 100 percent and your wealth increases instantaneously by a million dollars. Both *probability* and *wealth* are abstract mathematical quantities, like the wave function. You can't use

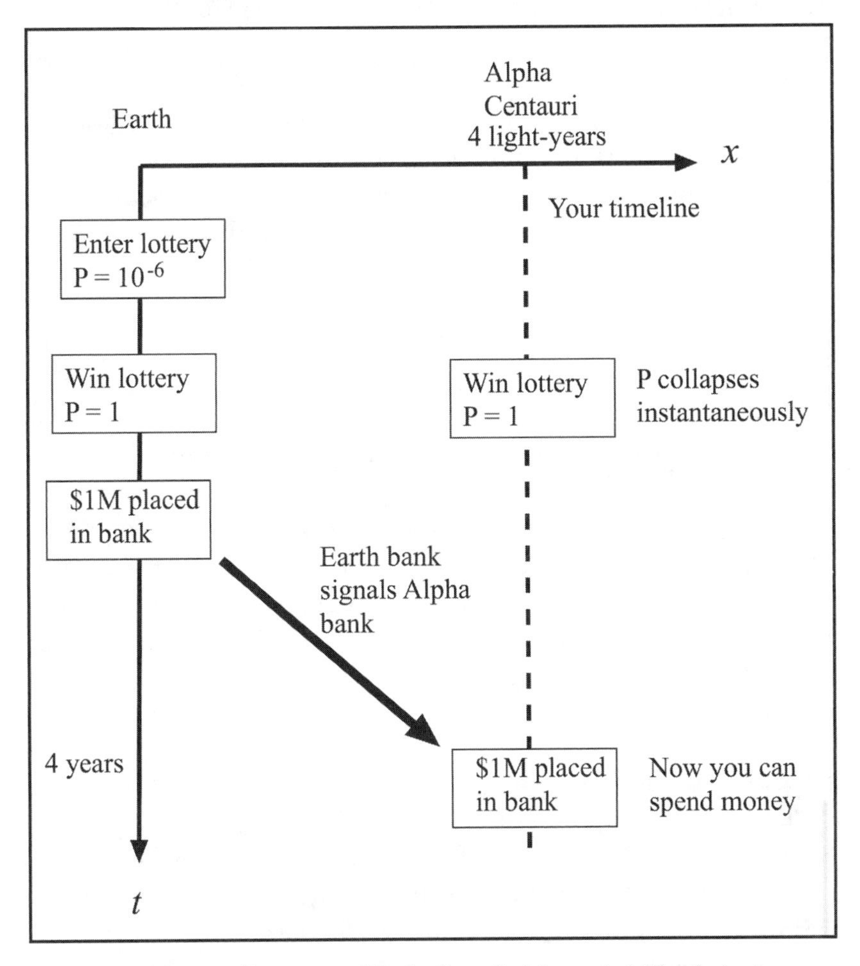

Fig. 12.4. An illustration of the "collapse" of the probability of winning a lottery on Earth while you are on a plant in Alpha Centauri. Although it goes from one in a million to one instantaneously, you can't spend the money until the signal from Earth reaches your bank four years later.

them for real cash. It takes four years for the news, traveling at the speed of light, to reach you and your Alpha Centauri bank. You can't start spending the money until that happens.

And that's how it is in conventional quantum physics. The collapse of the abstract wave function is just a mathematical artifice. As we will see in the next chapter, some interpretations of quantum mechanics do not involve wave function collapse. Even though this happens faster than the speed of light, relativity and the laws of conventional physics will limit any signal or other practical result.

Today, the wave function is itself no longer considered a necessary part of quantum mechanics. In Paul Dirac's classic textbook in which he presented the modern formalism for quantum mechanics, the term *wave function* appears only once—in a dismissive footnote, where he says: "The reason for this name [wave function] is that in the early days of quantum mechanics all the examples of these functions were in the form of waves. The name is not a descriptive one from the point of view of the modern general theory." Note that "modern" here refers to the year 1930.[19] Dirac's methods are still standard today.

In Dirac's quantum mechanics, the state of a system is represented by a vector in an abstract, multidimensional space. In Heisenberg's matrix mechanics, the state is an equally abstract mathematical object called a matrix (just a table of numbers). The wave function only appears in Schrödinger's wave mechanics, the least sophisticated quantum formalism. Nevertheless, with or without a wave function, the different versions of quantum mechanics yield the same statistical results, the same indeterminism with respect to individual systems.

THE QUANTUM BRAIN

In *The Unconscious Quantum* I presented a criterion for determining whether a system must be described by quantum mechanics: If the product of a typical mass (m), speed (v), and distance (d) for the particles of the system is on the order of Planck's constant (h) or less, then you cannot use classical mechanics to describe it but must use quantum mechanics.[20] Applying the criterion to the brain, I took the typical mass of a neural transmitter molecule ($m = 10^{-22}$ kilogram), its speed based thermal motion

(v = 10 meters per second), and the distance across the synapse ($d = 10^{-9}$ meter) and found mvd = 1700h, more than three orders of magnitude too large for quantum effects to be *necessarily* present. This makes it very unlikely that quantum mechanics plays any direct role in normal thought processing.

In chapter 2 I presented the proposal of Penrose and Hameroff that claims to overcome the fact that the basic interactions at the neuronal level are most likely classical. Hameroff was one of the subjects interviewed in *What the Bleep Do We Know!?* In his *Scientific American* column of January 2005, Michael Shermer gave *Bleep* a scathing review.[21] In discussing the possibility of a quantum brain, Shermer refers to my calculation. On his Web site, Hameroff scoffs, "I've not seen this proposal in a peer reviewed journal, nor listed anywhere as a serious interpretation of quantum mechanics."[22] Actually, my criterion is based on textbook quantum mechanics, originating with Niels Bohr in 1913—hardly in need of peer review.

Furthermore, in *The Unconscious Quantum* I make it clear that my criterion applies as a necessary condition, that is, when mvd is on the order of h or smaller, then you *must* use quantum mechanics. This is not the case for the brain. However, it is well known that macroscopic quantum systems such as lasers and superconductors exist that utilize quantum coherence, that is, quantum phenomena are still possible with $mvd >> h$, but only under very special situations.

But then Hameroff adds, "Nonetheless I agree with Stenger that synaptic chemical transmission between neurons is completely classical. The quantum computations we propose are isolated in microtubules *within* neurons." These supposedly utilize quantum coherence.

In a 1999 paper, physicist Max Tegmark looked at the problem of quantum coherence in the brain. Quantum phenomena are characterized by coherence effects, as exemplified by the double-slit experiment, where the wave functions of particles emerging from each slit maintain a constant phase relationship and interfere with one another.[23] When that phase relationship is destroyed, we have "decoherence" and the quantum effects go away. Tegmark determined that the decoherence timescales would be ten or more orders of magnitude shorter than the timescales for events in the brain. The brain is simply too large and too hot to be a quantum device. The brain seems to be a Newtonian machine and perhaps it evolved that way—to allow for a high level of predictability in our lives.

It is safe to say that the Penrose-Hameroff model has not been supported by the evidence to the satisfaction of the great majority of neuroscientists, including Jeffrey Satinover, author of *The Quantum Brain*, who was another *Bleep* contributor, as mentioned in chapter 2.[24] However, let us assume Penrose is right about the brain not being a strict algorithmic computer. A simple mechanism exists, well known to complexity theorists, which can enable the brain or an electronic circuit to act occasionally in a noncomputable way.

External sources in the environment such as cosmic rays or internal sources such as radioactive potassium (K40) in blood can be expected to induce fluctuations in brain currents. These processes are quantum in origin, which means that they are random—at least in most interpretations of quantum mechanics. Like the fluctuations that provide for mutations in the evolutionary process, these might serve to trigger what complexity theorists call a *bifurcation*, when a system moves from one quasi-stable state to another.

The brain could operate that way, being basically classical and deterministic, but occasionally being jolted by a random quantum event. What is interesting is that the decisions made on this fashion would be indistinguishable from creative acts or free will. Is that all there is to it?

NOTES

1. Amit Goswami, *The Self-Aware Universe* (New York: Penguin, 1993), p. 64.

2. Harold E. Putoff and Russell Targ, "A Perceptual Channel for Information Transfer over Kilometer Distances: Historical Perspective and Recent Research," *Proceedings of the IEEE* 64 (1976): 329–54.

3. Robert Jahn, "The Persistent Paradox of Psychic Phenomena: An Engineering Perspective," *Proceedings of the IEEE* 70 (1982): 135–70.

4. Goswami, *The Self-Aware Universe*, p. 131.

5. Victor J. Stenger, *Physics and Psychics: The Search for a World beyond the Senses* (Amherst, NY: Prometheus Books, 1995); *Has Science Found God? The Latest Results in the Search for Purpose in the Universe* (Amherst, NY: Prometheus Books, 2003), ch. 10.

6. Stenger, *Physics and Psychics*, p. 153.

7. Ibid., pp. 155–58.

8. W. B. Carpenter, "Spiritualism and Its Recent Converts," *Quarterly Review* 131: 301–53.

9. Ibid.

10. Ibid., pp. 166–74; C. E. M. Hansel, *The Search for Psychic Power: ESP and Parapsychology Revisited* (Amherst, NY: Prometheus Books, 1989).

11. Robert G. Jahn, B. J. Dunne, R. D. Nelson, Y. H. Dobbins, and G. J. Bradish, "Correlations of Random Binary Sequences with Pre-Selected Operator Intention: A Review of a 12-Year Program," *Journal of Scientific Exploration* 11, no. 3 (1997): 345–67. For my critiques, see Stenger, *Physics and Psychics*, pp. 181–84; *Has Science Found God?* pp. 281–85.

12. Stanley Jeffers, "PEAR Lab Closes, Ending Decades of Psychic Research," *Skeptical Inquirer* (May/June 2007): 16–17.

13. Dean Radin, *The Conscious Universe: The Scientific Truth of Psychic Phenomena* (New York: HarperCollins, 1997).

14. Douglas M. Stokes, "The Shrinking Filedrawer: On the Validity of Statistical Meta-Analysis in Parapsychology," *Skeptical Inquirer* 25, no. 3 (2001): 22–25.

15. Victor J. Stenger, *God: The Failed Hypothesis—How Science Shows That God Does Not Exist* (Amherst, NY: Prometheus Books, 2007), pp. 86–89.

16. Bela Scheiber and Carla Selby, eds., *Therapeutic Touch* (Amherst, NY: Prometheus Books, 2000).

17. Richard P. Feynman, *QED: The Strange Theory of Light and Matter* (Princeton, NJ: Princeton University Press, 1985), p. 15.

18. Mathematically, we write $\psi = a + ib = A\exp(i\theta)$, where $i = \sqrt{-1}$.

19. Paul Dirac, *The Principles of Quantum Mechanics* (Oxford: Oxford University Press, 1930). This book has had four editions and at least twelve separate printings.

20. Stenger, *Physics and Psychics*, p. 284.

21. Michael Shermer, "Quantum Quackery," *Scientific American* 292, no. 1 (January 2005): 234.

22. Stuart Hameroff, "Hackery/Quackery" *Scientific American*, http://www.quantumconsciousness.org/hackery.htm (accessed May 28, 2008). As quoted in Alexandra Bruce, *Beyond the Bleep: The Definitive Unauthorized Guide to* What the Bleep Do We Know!? (New York: Disinformation Company, 2005), p. 77.

23. Max Tegmark, "The Importance of Quantum Decoherence in Brain Processes," *Physical Review E* 61 (1999): 4194–206.

24. Bruce, *Beyond the Bleep*, p. 77.

13

QUANTUM PHILOSOPHY

If one abandons the assumption that what exists in different parts of space has its own, independent, real existence, then I simply cannot see what it is that physics is meant to describe. For what is thought to be a "system" is, after all, just a convention, and I cannot see how one could divide the world objectively in such a way that one could make statements about parts of it.

—Albert Einstein[1]

INTERPRETATIONS

Since the early days of quantum mechanics, philosophers and philosophically oriented physicists, as many physicists were then in contrast to today, debated the meaning of quantum mechanics. This story has been discussed in many books, including my own 1995 effort called *The Unconscious Quantum*, which goes into many details that I will not repeat here.[2] For my purposes in this book, I need review only the main ideas of the various interpretations and highlight the elements thought to be spooky.

Copenhagen Positivism

The *Copenhagen interpretation* was proposed by Niels Bohr in the late 1920s with input from Werner Heisenberg and others. They put into practice the philosophical doctrine called *positivism* that was mentioned earlier. Originally proposed in the nineteenth century by the philosopher Auguste Comte and the physicist and philosopher Ernst Mach, positivism asserts that knowledge can only be obtained by strict scientific method. According to this view, what we measure in scientific experiments and observations is all we know and all we can know about whatever reality is out there.

For a brief period early in the twentieth century a variation called *logical positivism*, or *logical empiricism*, was investigated by a group of philosophers in Europe calling themselves the *Vienna Circle*. These included the eminent philosophers Moritz Schlick, Otto Neutrath, Alfred Jules Ayer, and Rudolph Carnap, along with physicist Philipp Frank.[3] The Vienna Circle attempted to develop a standard language of knowledge that was based on empirical data alone, an idea proposed by Ludwig Wittgenstein that he later repudiated.[4]

Einstein had applied positivist thinking in his theory of special relativity by defining time as what you read on a clock and distance as what you read off a meter stick. Although he later disassociated himself from the doctrine, positivism makes relativistic effects that violate common sense, such as time dilation and the Fitzgerald-Lorentz contraction, almost trivial to understand. That is just how clocks and meter sticks behave by definition.

Recall the light-pulse clock in figure 8.3 (p. 112). The tick rate slows down when the clock is seen to be moving. If time is what you read on a clock, then it depends on your frame of reference. Now, you might argue that this has been shown only for the light-pulse clock. However, if only light-pulse clocks and not wristwatches or the body clocks of humans did not all tick at the same rate when in the same frame of reference, time would not be a very useful concept. In particular, it would violate the principle of Galilean relativity, discussed in chapter 6, which says that there is no difference between being at rest and moving at constant velocity. If clocks in a given reference frame read differently depending on whether or not the reference frame was moving, Galileo's principle would be violated because you could then have a way to distinguish being at rest from moving. In any case, theory aside, by now time dilation is a well-established empirical fact—even at an everyday speed such as that of a modern jetliner, where the effect is very tiny but measurable with atomic clocks.

In chapter 5 space and time were defined by what you ultimately measure on a clock. Distance is the time it takes light to go from one point in space to another in a vacuum. All physical quantities reduce to measurements off dials of some sort, which reduce to clock measurements. Of course we use meter sticks, thermometers, barometers, voltmeters, and other instruments to measure these quantities, but fundamentally they are calibrated against the standard clock.

Bohr and Heisenberg went much further than Einstein in their application of positivism. They proposed that the properties of an object themselves are determined by what you measure. When you measure an object's position you are not measuring something that is already there—you are giving it that position! When you measure an object's wavelength, or what Louis de Broglie showed is inversely proportional, its momentum, you are giving it that wavelength or momentum! That is not to say that these quantities are arbitrary. The experimenter does not decide what the values of these measurements are. The values observed are not certain but follow a probability distribution that is determined by whatever objective reality is out there.

In the Copenhagen interpretation, the path of a particle is not determined until it is measured. Thus, unless a detector is placed in the path, as in figure 8.6 (p. 120), all paths are possible and so they can interfere. This interference comes about because the state vector must contain all possibilities and so is an "entangled" mixture of the state vectors for each path.

According to Copenhagen, the result of a measurement is generally not fully predetermined. As mentioned above, it can have some value chosen from a probability distribution that is given by the wave function. For example, suppose that an observation has two possible outcomes, **A** with a probability of 75 percent, and **B** with a probability of 25 percent. We cannot predict the outcome of any given measurement, just that in a large ensemble of identical measurements we will get, on average, three times as many results **A** as **B**. The act of measurement that is represented by wave function collapse in the Copenhagen interpretation (see chapter 8) is not included in the formalism and left somewhat mysterious.

Einstein and Bohr held a long-standing, collegial debate over quantum mechanics that never resulted in final agreement.[5] The majority of physicists ultimately accepted Bohr's views for a good thirty years. Today, no consensus exists on the "best" interpretation, although Copenhagen is generally regarded as outdated.

Still, the quantum spiritualists take Copenhagen as a basis for their claim that the mind controls reality since it is the mind that decides what and when to measure and thus collapses the wave function. Since collapse happens throughout the universe and back in time, the mind must be tuned in to some holistic "cosmic consciousness."

This was not Bohr's view. He never attributed an active role for consciousness in the measurement process. It was simply part of what he called the "observer" that could just as well be a passive measuring instrument like a Geiger counter as a human investigator. Bohr's observer was treated as a separate, macroscopic system outside the quantum system being observed.[6] And when he said that the act of measurement gave an object the property being measured, he was not claiming some kind of mental control over reality. Those properties were simply defined by their measurement just as time is defined by what you read on a clock.

However, other prominent physicists gave consciousness a larger role. John von Neumann conceived of the observational act as a sequence of processes in which information is transferred from the object to the observer's consciousness.[7] Eugene Wigner asserted that consciousness or our mind is able to affect matter. He is quoted as saying, "The laws of quantum mechanics cannot be formulated . . . without recourse to the concept of consciousness."[8]

Today, positivism has been discarded as philosophers admit that some objective reality exists beyond the acts of human observation. However, observations are still our only reliable means for learning about that reality and measured quantities such as space and time need not exist in one-to-one correspondence to that objective reality.

Hidden Variables

Despite the dominance of Bohr and the Copenhagen interpretation, physicists had long sought a subquantum theory that was more like conventional Newtonian physics—deterministic, easier to understand, and less spooky. The instantaneous collapse of the wave function certainly added to the spookiness. In chapter 8 we mentioned how de Broglie's proposal in 1927 that the wave function is a "pilot wave" guiding a particle's motion was ignored. The quantum physics consensus at the time was so dominated by Bohr that even the great Einstein could not break through, although he and two collaborators did plant a few philosophical doubts in 1935 with the EPR paradox.

In 1952 David Bohm produced a variation on de Broglie's pilot waves and showed that the notion of subquantum forces is possible. This is usually called the *hidden variables* interpretation of quantum mechanics, although Bohm preferred *ontological interpretation.*[9] As we saw in chapter 8, this proved testable against conventional, statistical quantum mechanics and the test results agreed precisely with convention, ruling out any theory of *local* hidden variables.

Bohm did not discard his proposal, simply noting that the hidden variables must be nonlocal, that is, act over distances faster than the speed of light. Furthermore, his nonlocal theory provided him with a basis for a metaphysical model in which the universe is one interconnected whole.

However, Bohm's nonlocal theory has several strikes against it that have prevented it from being ranked very high among physicists as the interpretation of choice. First, it only provided another way to look at the Schrödinger equation, offering nothing new in calculational ability and producing no unique empirical results. Second, it dealt, as does the Schrödinger equation, with only nonrelativistic particles—particles moving slowly compared to the speed of light. A huge amount of quantum physics has gone over the bridge since Schrödinger introduced his equation in 1926 and Bohm's theory has nothing to say about any of that physics. Third, it implies superluminal connections, in violation of Einstein's speed limit.

Many Worlds

In the Copenhagen interpretation, a quantum system is observed from outside the system, with the observing apparatus treated classically. Furthermore, no mechanism is provided for the act of measurement itself, which is taken as being randomly chosen from the statistical distribution given by the wave function.

In his PhD thesis from Princeton University in 1957, titled "The Relative State Formulation of Quantum Mechanics," Hugh Everett III rectified that problem by developing a mathematical formalism for quantum mechanics that treated the measuring instruments and observers as all part of a single quantum system along with the system being observed.[10] Everett's approach was later dubbed more grandly by Bryce DeWitt as the *many-worlds interpretation* (MWI) of quantum mechanics.[11]

Not only did Everett solve the problem of treating the observer and the observed as part of a single quantum system, he accounted for the statistical distribution of outcomes.

Consider the example used above in which a measurement has two outcomes **A** and **B** with relative probabilities 3:1, respectively. Prior to a measurement the universe is viewed as following a path through an abstract space. Upon the act of measurement the path splits into four, three corresponding to outcome **A** and single path corresponding to outcome **B**. This is illustrated in figure 13.1. The statistical distribution of measurements is accounted for by having three times as many paths with outcome **A** than have outcome **B**. The particular outcome **A** will be obtained on average every three out of four times, with outcome **B** occurring on average once in four.

DeWitt and others interpreted Everett's formulation as implying that upon the act of measurement the universe splits into different universes or "worlds," depending on the outcome of the experiment. In the example

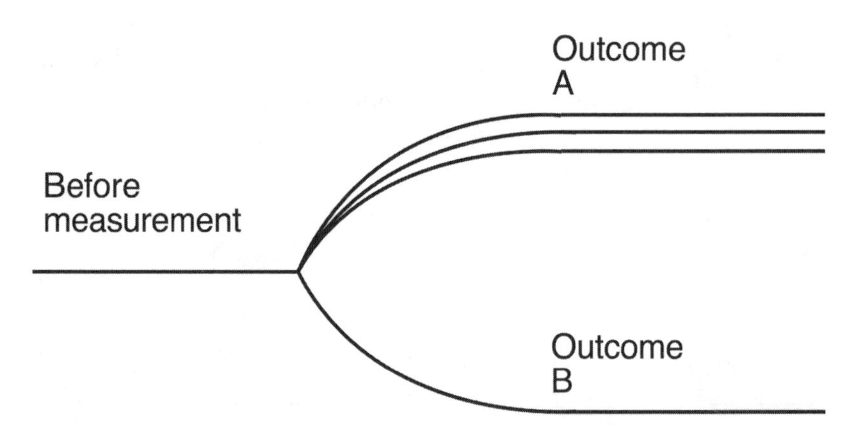

Fig. 13.1. The many-worlds interpretation viewed as different paths through an abstract space. The act of measurement is described as causing the path to split into equal probable paths, where the number of paths for each outcome reflects the probability of that outcome. In the conventional MWI, each path occurs in a separate parallel universe.[12]

above, we have three worlds in which the result of the measurement is **A** and another world in which the result is **B**. All possible worlds exist and all possible events take place somewhere. You and I live in one of the worlds with a particular series of measurement outcomes.

MWI helps us understand two-slit interference without the introduction of the wave-particle duality. In the particle picture, the photon or the electron must pass through one slit or the other, so there is no particle passing simultaneously through the other slit with which to interfere. In MWI, there is one world where the particle passes through one slit and another world where it passes through the second slit. As Everett found, the wave function of the detector is "entangled," containing a piece for each world, and so the result of the measurement exhibits the interference between the two.

This example emphasizes a point not generally recognized: the dif-

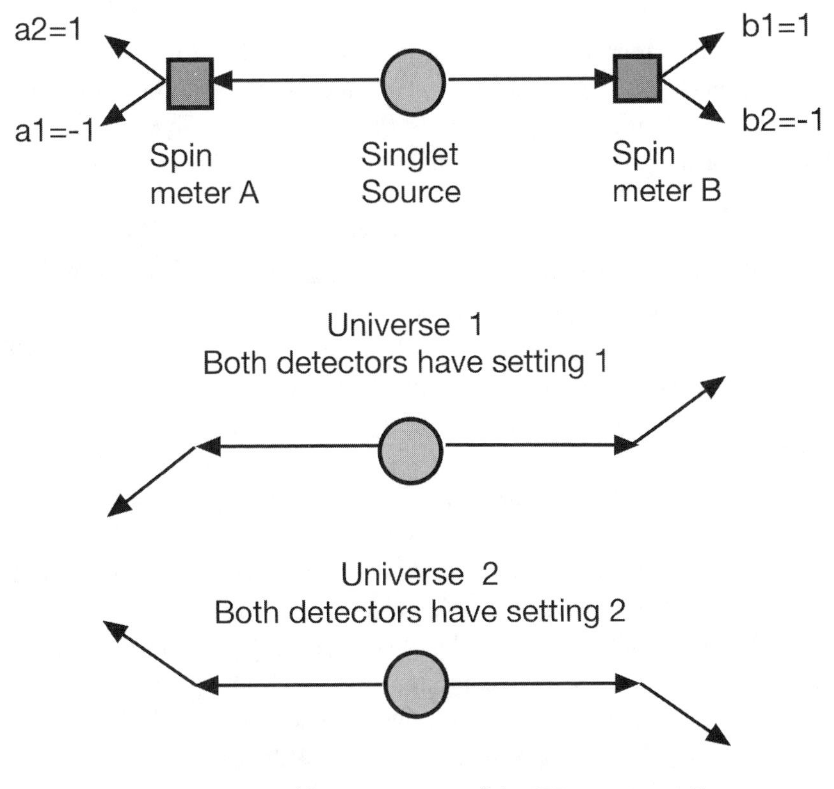

Fig. 13.2. Many-worlds interpretation of the EPR experiment. Two universes exist, one for each possible pair of spin meter settings.

ferent worlds in MWI are not independent of one another. They all connect at the point of measurement. Physicist David Deutsch, one of the strong proponents of MWI, takes the phenomenon of interference as evidence that the model describes objective reality as it really is. He says, "We do not need deep theories to tell us that parallel universes exist—single particle interference phenomena tell us that."[13]

The many-worlds interpretation also gives us a simple explanation for the results of the EPR experiment. As shown in the figure, there are two settings numbered 1 and 2, where the axes along which the spins are measured are opposite. The detector at the end of one beam line seems to "know" what the setting is at the end of the other beam line so that the total spin comes out zero. This requires a superluminal signal since the setting can be done after the photons have left the source. In MWI there are two universes, one for each setting.

Although MWI has received wide recognition as a result of Everett's elegant mathematical description, most physicists and philosophers are not quite ready to go as far as Deutsch in granting it recognition as the only correct interpretation of reality rather than simply a mathematical representation of that reality. They are simply uncomfortable with the idea of a limitless number of parallel universes.[14]

Furthermore, as we saw above for hidden variables theories, before most physicists will agree to anoint a theory as the consensus view they expect to see some empirical tests favoring that theory over all its competitors. In the case of MWI and any of the other viable interpretations of quantum mechanics, all yield the same observations. (Those interpretations that disagree with the data are falsified and so are not viable.)

Despite being the most mysterious of any of the interpretations of quantum mechanics, quantum spiritualists do not find MWI supportive of their claims—except in a vague version called *many minds* in which each world is a separate mind. Notice that MWI, however, does make possible a deist god who has completely predetermined everything that happens in the multiverse of all-possible worlds. Those in any given world see things statistically, but this god is looking down on the whole picture. In figure 13.1, for example, he sees all four paths. The problem of producing humanity is solved since one set of paths will likely lead that way.

However, human free will as we conceive it does not exist in many-worlds quantum mechanics. What we have is the appearance of free will in each of the different worlds, where chance seems to decide which path is followed. For example, suppose a man is looking at the menu in a restaurant and can't decide between steak and lobster. God's plan has him eating steak three times as frequently as lobster. Referring to figure 13.1, the top three branches have him eating steak and the lower branch has him eating lobster. We thus have four worlds, one with a man eating steak in three worlds and lobster in one. In each world the situation appears as if the man made a free choice. But the deist god had it all figured out ahead of time; at a particular point in space-time a man in four different worlds would eat steak in three and lobster in one.

Of course this solution to the free will problem does not satisfy the Christian need for sin and redemption. Humans should not be punished when their choices were really predetermined after all.

Histories

Let us look at some of the later attempts to achieve an interpretation of quantum mechanics that has all the features of Everett's formalism without the baggage of parallel worlds. In 1984 Robert Griffiths provided a sound logical and mathematical interpretation of quantum mechanics called *consistent histories.*[15] He considered all the possible paths of a quantum system and postulated only those occur that are logically consistent in the sense that they enable you to calculate a conventional probability for those paths. While this interpretation works in the sense that it gives all the results of quantum mechanics, it does not provide us what we would like to have, namely, some kind of intuitive picture of what is happening in reality.

In 1990, Murray Gell-Mann and James Hartle elaborated a related interpretation they called *alternate histories.* They provide an explanation for why only Griffith's consistent histories are recorded by our measuring apparatus: "An 'observer' (or information gathering and utilizing system) is a complex adaptive system that has evolved to exploit the relative predictability of a semi-classical domain."[16] That is, it is to our evolutionary advantage to be able to predict phenomena with some confidence so we have adapted to follow those sets of consistent histories that relate a classical world with a predictive causal structure. However, it is not clear to me how we do this. How do we decide what path to follow? Furthermore, recent studies indicate that there are many consistent histories that do not give classical outcomes.

Decoherence

Finally, let me mention the idea of *decoherence*, which provides a mechanism that at least mimics wave function collapse.[17] Two waves are said to be *coherent* when they have the same frequency and a constant phase difference. Coherent waves produce interference patterns. (The "entangled" states we talked about earlier are coherent states.) When the phase difference between waves is random, they are said to be *decoherent* and no interference is seen. A wave that scatters off some object in the environment will usually undergo a random change in phase. That scattering object could be a particle detector. Thus, the detection of a particle, or its interaction with the environment, not only specifies the path of that particle but also randomly changes the phase of the wave function of a particle so that it is no longer "coherent" with the rest of the system.

For example, the probability for a photon in the visible region of the electromagnetic spectrum interacting with an air molecule is low so that a beam of visible light can remain coherent, that is, able to produce interference and diffraction effects, over a distance of kilometers. That is, it looks to us like a wave. On the other hand, a high-energy photon called a *gamma ray* has a high probability of interacting and will decohere over a short distance. A beam of gamma rays will then appear to us as a beam of particles and produce no interference or diffraction effects. The same is true for electrons and other "particles" as normally observed, which is why we generally identify them as particles while we identify visible light as a wave.

Decoherence explains why objects on the macroscale do not normally exhibit interference and diffraction—we do not see them bend around corners. It is because they cannot be effectively isolated from their environment and so quickly decohere. We saw this was important in chapter 12 where we discussed whether the brain takes advantage of quantum mechanics in creating consciousness.

Recall the discussion in chapter 8 of the double-slit experiment with a detector placed outside one slit, as illustrated in figure 8.6 (p. 120). There I said, "Let us assume for the purposes of this discussion that we can detect the electron without seriously deflecting it from its path or taking away significant amounts of its energy." This was an unjustified assumption. To be detected the particle must be interfered with, which means the phase of its wave function is changed. Decoherence thus simply explains the apparent paradox of double-slit interference that I witnessed Richard Feynman talk about at Hughes Aircraft Company fifty years ago.

FEYNMAN PATHS AND TIME REVERSAL

All of these various ways of looking at quantum mechanics are expressed in terms of "paths." They all grow out of the brilliant alternative way of doing quantum mechanics that was invented by Feynman prior to World War II while in graduate school at Princeton and published as his PhD thesis.[18] Feynman showed that he could derive standard quantum mechanical probabilities by considering all the possible paths of a system from some initial point **a** to some final point **b**. The sum of the probability amplitude over those paths gave the probability amplitude for the system to go from **a** to **b**. The probability amplitude is a complex number. Squaring its magnitude gives the probability for the event. For example, in the double-slit experiment you have two paths from the source to a given point on the screen. Adding their amplitudes and calculating the probability gives the observed interference pattern. Feynman viewed this as simply a mathematical trick since the universe was assumed to follow only one path. In the many-worlds view each path is followed in a different world.

Recall that in chapter 5 we found that there is no fundamental direction to time. In *Timeless Reality* I showed how by means of *time reversal* we can view all of the Feynman paths as actually taking place—just as the mathematics suggests. Of course, MWI does this too, but with time reversal we have it all happening in a single universe.

Let us see how that happens in the case of the double-slit experiment. Refer to figure 13.3. The photon or the electron from the source goes forward in time through the top slit, then backward in time through the bottom slit. We thus have two paths and summing their amplitudes gives the interference pattern as before.

In chapter 9 we discussed how elementary particle interactions are described by means of *Feynman diagrams*. They are used to calculate the probability amplitude for the reaction taking place by that mechanism.

Again here we find that, as with the case of Feynman paths, a single diagram does not describe a specific event but one must sum the amplitudes over all possible diagrams (see figure 9.1). Time reversibility allows us to imagine how all the diagrams actually participate in a single event. The process given in one diagram takes place in one time direction, then another in the reverse time direction, then a third, and so on.

The Feynman diagram can be used to show how time reversibility makes it possible for a particle to be in two or more places at the same time without involving any superluminal motion.

Consider figure 13.4. In (a), a single electron goes forward in time, then back, and then forward, appearing simultaneously at three different positions in space at time **B**. The electron never moves faster than the speed of light.

Physicists conventionally assume a single direction of time, in which case, as Feynman first showed, the backward-moving electron is empirically indistinguishable from an antielectron, or a positron. The process is

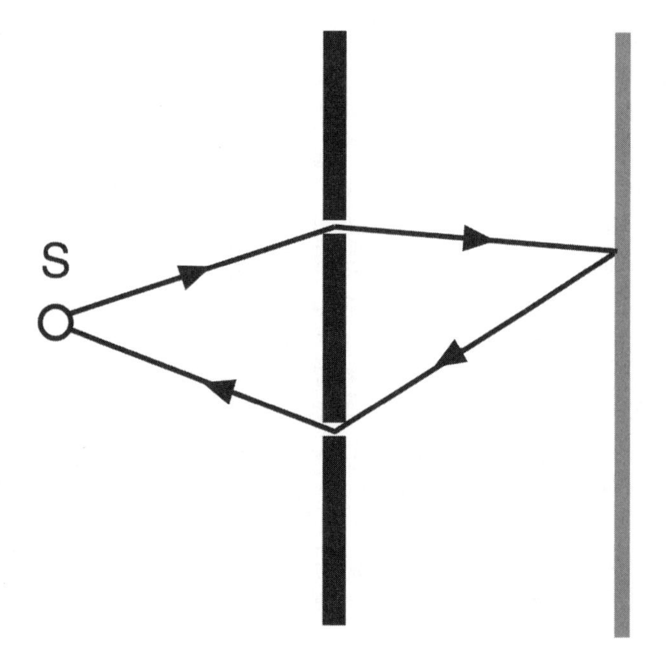

Fig. 13.3. The time-reversed double-slit experiment. The photon or the electron goes forward in time through one slit and backward in time to the source through the other slit.

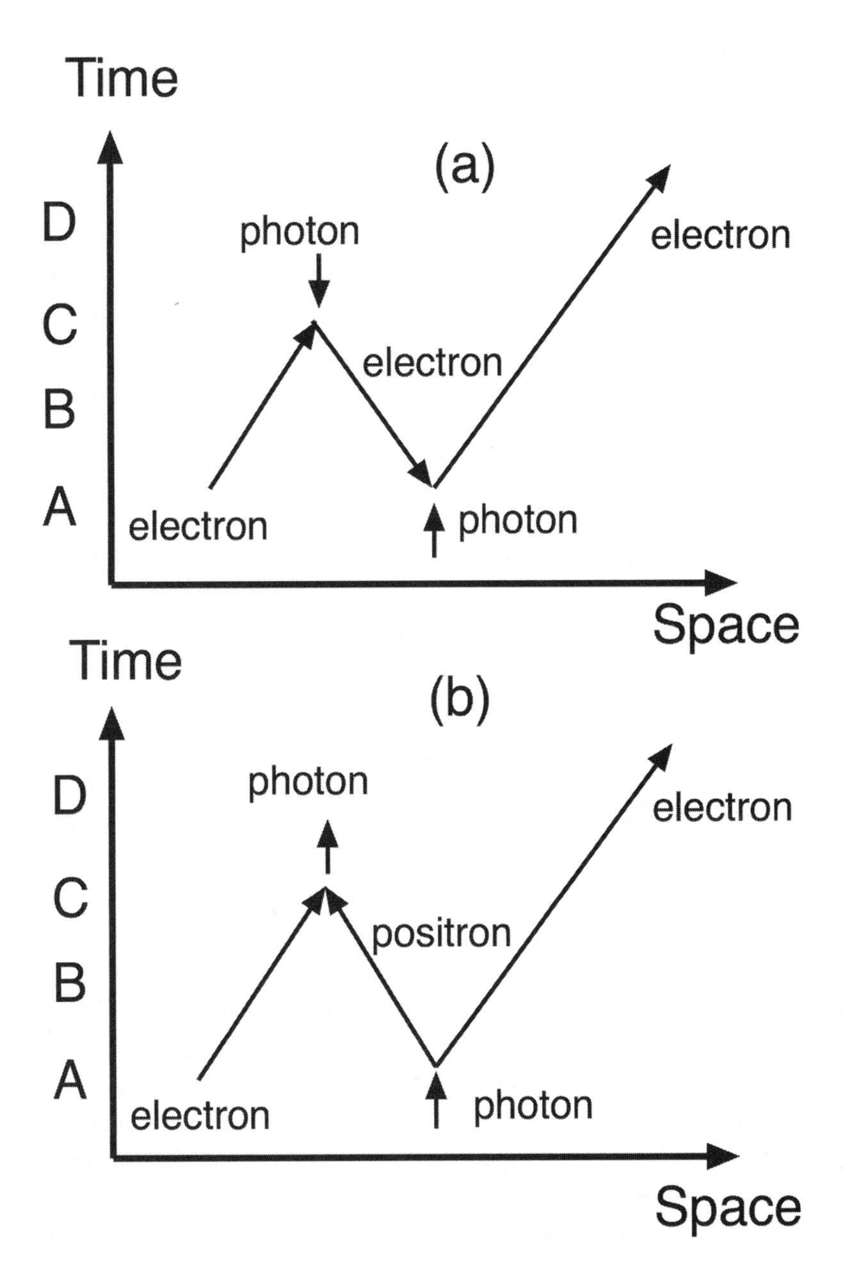

Fig. 13.4. In (a) we show an electron going forward, backward, and again forward in time so that at time **B** it appears simultaneously at three different positions in space. In (b) we see the convention interpretation assuming a single time direction. This requires the introduction of antiparticles.

then interpreted, as shown in (b), as an electron-positron pair being produced by a photon at time **A**, the positron of the pair then colliding with a original electron at time **C**, giving off a photon. Thus the electron at **D** is different from the electron at **B**.

The conventional view is clearly more complicated than the time-reversed view, with the need for three particles in the place of one. Furthermore, the time-reversible picture helps us understand why all elementary particles of a given type are indistinguishable from one another. They are the same particle!

DETERMINISM OR INDETERMINISM?

The only deterministic quantum theory is that of Bohm. However, as we saw above, the results of the EPR experiment show that local hidden variables, that is, hidden variables whose effects travel through space no faster than the speed of light, are ruled out by experiment.

This leaves us with two possibilities:

1. Indeterministic quantum mechanics in which only the average behavior of an ensemble of systems can be predicted, and any nonlocal effects are confined to the mathematics and not measurable quantities.
2. Deterministic quantum mechanics in which the forces that determine the motion of individual systems are necessarily nonlocal.

These two are, as far as we know, empirically indistinguishable. Bohm's quantum mechanics still gives the same statistical results as conventional quantum mechanics. Furthermore, Bohm's theory is incapable of dealing with relativistic particles, that is, particles moving at such high speeds (but still less than the speed of light), that relativistic effects must be considered. Thus it cannot be easily extended to relativistic quantum field theory, the basis of the standard model of particles and forces. This is rather ironic since Bohm's theory claims to describe connections that move faster than light but it can't handle particles moving at 90 percent of the speed of light.

In any case, the final test as it always is in science is what the data say.

No one has ever seen a particle moving faster than light nor transmitted information from one point to another superluminally. Furthermore, as we have already seen, the attempt to develop a holistic theory of particle physics has failed, while the traditional reductionist model of elementary constituents of matter has enjoyed continued success. Given all this, I think we can safely discard the quantum spiritualist notion of a holistic universe.

NOTES

1. Max Born, *The Born-Einstein Letters*, trans. Irene Born (New York: Walker and Co., 1971).

2. Victor J. Stenger, *The Unconscious Quantum: Metaphysics in Modern Physics and Cosmology* (Amherst, NY: Prometheus Books, 1995).

3. Alfred Jules Ayer, *Logical Positivism* (New York: Free Press, 1959).

4. Ludwig Wittgenstein, *Tractatus Logico-Philosophicus*, trans. by C. K. Ogden (Routledge and Kegan Paul, 1922); *Philosophical Investigations* (New York: McMillan, 1953).

5. Stenger, *The Unconscious Quantum*, pp. 66–79.

6. Willem M. de Muynick, *Foundations of Quantum Mechanics: An Empiricist Approach* (New York: Springer, 2002), p. 223.

7. Ibid., p. 224.

8. Eugene P. Wigner, "The Probability of the Existence of a Self-Reproducing Unit," in Michael Polanyi, *The Logic of Personal Knowledge* (Glencoe, IL: Free Press, 1961), p. 232.

9. David Bohm and Basil J. Hiley, *The Undivided Universe: An Ontological Interpretations of Quantum Mechanics* (London, New York: Routledge, 1993).

10. Hugh Everett III, "Relative State Formulation of Quantum Mechanics," *Reviews of Modern Physics* 29 (1957): 454–62.

11. Bryce S. DeWitt and Neill Graham, eds., *The Many-Worlds Interpretation of Quantum Mechanics* (Princeton, NJ: Princeton University Press, 1973).

12. One objection that has been raised is, what if the probabilities are irrational numbers like $1/\sqrt{2}$? I am assuming the frequency interpretation in which probabilities are like chances, 1 in 3, 400 in 500, and so on.

13. David Deutsch, *The Fabric of Reality: The Science of Parallel Universes—and Its Implications* (New York: Allen Lane, 1997), p. 51.

14. There is some dispute about how many physicists "believe" in the many-worlds interpretation of quantum mechanics. It all depends on how you ask the question. If you ask if they believe the Everett formalism is correct, most will say

yes. If you ask if they believe the universe splits in two with every blink of any eye, most will look at you as if you are crazy.

15. Robert J. Griffiths, "Consistent Histories and the Interpretation of Quantum Mechanics," *Journal of Statistical Physics* 26 (1984): 219–72.

16. Murray Gell-Mann and James P. Hartle, "Time Symmetry and Asymmetry in Quantum Mechanics and Quantum Cosmology," in *Proceedings of the 1st International A. D. Sakarov Conference on Physics, Moscow, May 27–31, 1991* and *Proceedings of the NATO Workshop on the Physical Origin of Time Asymmetry, Mazagon, Spain, September 30–October 4, 1991*, ed. by J. Haliwell, J. Perez-Mercader, and W. Zurek (Cambridge: Cambridge University Press, 1992).

17. Wojciech H. Zurek, "Decoherence, Einselection, and the Quantum Origins of the Classical," *Reviews of Modern Physics* 75 (2003): 715.

18. Richard P. Feynman, "The Principle of Least Action in Quantum Mechanics," PhD diss. (University Microfilms pub. no. 2948, 1942); R. P. Feynman and A. R. Hibbs, *Quantum Mechanics and Path Integrals* (New York: McGraw-Hill, 1965).

14

WHERE CAN GOD ACT?

I found extraordinary difficulty, when I thought about events in scientific terms, in imagining any kind of loophole through which God could influence them.

—William Pollard[1]

THE DEMISE OF ENLIGHTENMENT DEISM

We have seen how Newtonian mechanics implies that the laws of physics predetermine everything that happens in the material universe. It follows that if a creator god exists, he has nothing to do once he creates the universe and its laws and sets the initial conditions from which those laws take off.

This characterizes an impersonal *deist* god, a creator who does not act in the universe, as opposed to the personal *theist* God who continues to interact with the universe and its inhabitants after creation. Deism became the religion of many important intellectual figures in the Age of Enlightenment, when reason mounted its first serious challenge to scriptural and traditional authority in the history of Christendom. The primary founders of the American republic were desists including, it seems, the first four presidents.

Both the Enlightenment and deism faded early in the nineteenth century, as Europe endured the bloody French Revolution followed by the Napoleonic Wars, while the United States experienced the second phase of the so-called Great Awakening to a spiritual faith that placed feeling and emotion ahead of reason.

In the early twentieth century the rug was pulled out from under Enlightenment deism with the development of quantum mechanics. In Newtonian mechanics, the position and the momentum of a body are needed to predict its motion. The quantum Heisenberg uncertainty principle shows these cannot be measured simultaneously with unlimited precision. It follows that the motion of a body cannot be predicted with unlimited precision.

Note that the uncertainly principle did not eliminate the high predictability of physical events on the macroscale. Recall the example of a one-gram body initially located to within a cubic centimeter. The uncertainty in its speed is only 5×10^{-30} meter per second and so its motion using Newtonian mechanics can be predicted with a probability that, while not 100 percent, is surely sufficient for any practical purpose.

We have seen that the methods of quantum mechanics, as used in practice, only predict the statistical distribution of events and not the occurrence of any individual event. In most interpretations of quantum mechanics this is taken to mean that the events themselves are not predetermined intrinsically but happen by chance within limitations placed by global laws of physics such as energy and electric charge conservation. Note that the statistical distributions, as described by the quantum wave function are in fact predetermined by initial conditions and an equation of motion called the *time-dependent Schrödinger equation.* Still we can safely disregard the clockwork universe and with it Enlightenment deism.

Surprisingly, an unacknowledged deism seems to have remained in people's minds to the present day. We saw some evidence for this in chapter 1, where I reported on a survey that indicates some 44 percent of Americans believe in a god who does not act either in the universe or in their own lives. Although mostly professed Christians, they apparently do not hold the image of the traditional Christian God who steps in to alter the course of history. These ordinary laypeople have apparently intuited a fact that Christian theologians have finally begun to grasp: the Christian God is very difficult to reconcile with science, logic, or common sense. In place of theism a new kind of deism is being developed by some theologians and believing scientists, although they are not yet ready to admit that their new god has little in common with the traditional God of Christianity, Judaism, and Islam.

NO GOD IN THE GAPS

In this book I have not had much to say about the war between science and religion over the issue of evolution. That subject has been covered extensively in a host of other books. The opposition to evolution though often cast as a disagreement with mainstream science over alternative "theories" is motivated by the conviction, which I share, that evolution and biblical faith are irreconcilable. The difference I have with creationists is they think that the Bible is correct and science is wrong, while I think that science is correct and the Bible is wrong. As we will see, however, evolution does not conflict with the new deism.

The conflict between science and religion goes much deeper than creationism versus evolution. Evolution is just one component in the scientific worldview in which reality is composed solely of matter and nothing more—no spirits, souls, or gods. As I have shown in some detail, materialism is consistent with all scientific knowledge as well as commonplace experience. We have well-established theories that can be used to accurately describe most scientific observations as far out as we can see in space and as deep down as we can look into matter. We have no empirical fact that requires us to introduce anything beyond matter. While we must always remain open to the possibility that some new evidence will be found in the future that points to a spirit world, at this writing no such evidence exists.

Of course science does not know everything. We still have, and no doubt will always have, gaps in our scientific knowledge. Technically, this still leaves room for the immaterial or spiritual to appear—the so-called *God of the gaps*. However, the mere existence of a gap in knowledge cannot be used as an argument for the existence of some god or spirit, as long as we can give plausible tentative explanations that do not require the introduction of any immaterial or supernatural elements.

For example, while we do not know exactly the mechanism by which our universe appeared 13.7 billion years ago, we can present any number of plausible scenarios based on well-established physics and cosmology. I presented two such scenarios in my book *The Comprehensible Cosmos*.[2] They will be discussed in the last chapter.

Similarly, we cannot describe exactly how life originated, but many proposed scenarios consistent with well-established chemistry and biology can be found in reputable scientific journals.[3] Thus no rational basis exists for claiming that a supernatural origin for life or the universe must have occurred. The same is true across the board: from cosmology to neuroscience, no case can be made that we need something more than matter to understand the universe.

THE PREMISE KEEPERS

Many contemporary Christian theologians and theistic scientists accept the results of science and do not dispute the power of its meticulous procedures. Nevertheless, they still assume that a world beyond matter exists. They argue that religious belief has been so persistent throughout history that there has to be something to it. They then proceed to make an honest attempt to reconcile science with the images of God drawn from traditional beliefs.

In earlier writings, which mainly focused on evolution theology, I referred to this group as the "premise keepers."[4] They include, among others, the particle physicist and Anglican priest John Polkinghorne, the biochemist and Anglican priest Arthur Peacocke, the evolutionary biologist and devout Catholic Kenneth Miller, the physicist and theologian Ian Barbour, the cosmologist and Quaker George Ellis, the physicist and theologian Willem Drees, and theologians John Haught and Nancey Murphy. Here I will review their attempts to find a way for God to act in the world and ask whether they are viable in the light of modern science.

The problem of locating God's action was the subject of a multiyear collaborative project between the Vatican Observatory and the Center for Theology and the Natural Sciences headquartered in Berkeley, which alone testifies to the fact that this is not a settled matter even in the Catholic Church. Five volumes of proceedings edited by center director Robert John Russell and various other scholars were produced.[5] In 2006 a whole issue of *Zygon, the Journal of Religion and Science* was devoted to the question.[6] Comprehensive analyses of divine action can be found in the book by Nicholas Saunders[7] (see also Saunders's article in *Zygon*) and the review articles by Wesley J. Wildman.[8] A number of other books on the subject of varying scholarly quality have also been published.[9]

THE VATICAN SERIES

In this section I will relate some of the key arguments made in first three volumes of the Vatican series (the fourth was not available at this writing). Here I will be relying mainly on material from the overview on the project's Web site. Looking at the originals I am confident that these summaries are accurate and of excellent quality. The outer quotations ("summary quotation") are taken from those summaries; the inner quotations ('author quotation') are from the original authors. My own comments are occasionally added.

It is not possible for me to cover every article in the series and I have selected those that I feel provide a good representation of the range of views presented. Also, keep in mind that, independent of their own personal beliefs, all the authors are writing from a theological perspective in which God is assumed to exist and the question is what role he plays in the universe.

VOL. 1. *QUANTUM COSMOLOGY AND THE LAWS OF NATURE* (1993)

William Alston, "Divine Action, Human Freedom, and the Laws of Nature"

Alston argues that because of quantum indeterminism, God can act without violating physical law. However, he points out that even when you have only deterministic laws, they only allow absolute predictions for *closed* systems, that is, systems that have no outside influences. We have no way of knowing if the universe is closed and we know all the operative forces at work. "Hence, in this more general sense, God's acts do not violate natural law regardless of whether these laws are probabilistic or deterministic."

Comment: Even if we can't make predictions in an open deterministic universe (and in many cases we can), it is still deterministic, so God can't act unless he breaks his own laws. Besides, the universe is a closed system unless there is some outside force. What force can that be except some kind of god?

Paul C. W. Davies, "The Intelligibility of Nature"

"What is most significant about nature is that the universe is '...poised, interestingly, between the twin extremes of boring over-regimented uniformity and random chaos.' Accordingly it achieves an evolution of novel structures through self-organizing complexity. 'The laws...encourage physical systems to self-organize to the point where mind emerges from matter, and they are of a form which is apprehendable by the very minds which these laws have enabled nature to produce.'" However, Davies does not claim that this leads to an argument for God. He prefers "an evolutionary interpretation of mind as emergent within the material process of self-organization. The emergence of mind with its ability to pursue science is not just a 'biological accident.' Instead it is inevitable because of the laws of physics and the initial conditions. Hence life should emerge elsewhere in the universe—a claim which Davies sees as testable."

Comment: This supplements our discussion of emergence in chapter 10. We will later discuss the role of emergence in theology.

Thomas F. Tracy, "Creation, Providence, and Quantum Chance"

"Theologians from deists to liberals such as [Friedrich] Schleiermacher, [Rudolf] Bultmann, and [Gordon] Kaufman, have worked with a closed causal picture of the world that they feel is authorized by science. They have taken this to be incompatible with divine action in the world, leaving either a God who only sets the world's initial conditions or whose actions violate the laws of nature. But contemporary natural science does not necessarily lead to a deterministic metaphysics. Tracy cites two possible responses. First, a theologically sufficient account of God's particular actions in history might actually be developed that still limits God to being the creator of history as a whole. Second, God can be said to act in particular cases without intervention in history if one can defend an indeterministic interpretation of natural causes. It is here that quantum physics might be relevant."

Comment: This sounds like the deist god, although Tracy allows action in "particular cases" that do not change history. Those actions can't be very important, then.

VOL. 2. *CHAOS AND COMPLEXITY* (1996)

James P. Crutchfield, J. Doyne Farmer, Norman H. Packard, and Robert S. Shaw, "Chaos"

"The amplification of small fluctuations may be one way in which nature gains 'access to novelty' and may be related to our experience of consciousness and free will."

Willem B. Drees, "Gaps for God?"

"Theories of chaotic and complex systems have made it clearer than ever before that a naturalistic explanation of the world is possible, even in light of the lack of predictability of these systems. These theories have effectively closed certain gaps in our understanding of nature. [Drees is] critical of John Polkinghorne's suggestion that the unpredictability of natural processes provides a potential locus for divine action. Polkinghorne suggests that God brings about an input of information into the world without an input of energy. Drees claims that this is inconsistent with quantum physics and thermodynamics. In addition, Polkinghorne seems to interpret the unpredictability of chaotic systems as a sign of intrinsic openness, but this ignores the real meaning of deterministic chaos. Moreover, discarding the theory of deterministic chaos would be inconsistent with the very critical realism that Polkinghorne promotes."

Comment: Drees is the most atheistic Christian theologian I know of.

George F. Ellis, "Ordinary and Extraordinary Divine Action: The Nexus of Interaction"

"Some account of *special* divine action is necessary if the Christian tradition is to make sense. However, there are two important constraints to be reckoned with. One is that an ideal account of divine action must not conflict with a scientific understanding of nature; the other is that some explanation must be given of why a God capable of special action would not exercise that ability regularly to oppose evil and ameliorate suffering.

"[Ellis's] analysis of top-down causation convinces him that this concept alone does not provide for an adequate account of divine action.... A study of the possibilities for divine action via top-down causation leads inevitably to a consideration of divine action at the quantum level.

"Ellis takes God's action to be largely *through* the ordinary created processes. God initiates the laws of physics, establishes the initial conditions for the universe, and sustains the universe and its processes, which in turn result in the emergence of higher levels of order, including, finally, free human beings. Special divine action focuses on providing to human beings intimations of God's will for their social lives. Thus, the problem of the mode of divine action is largely a question of how God might communicate directly with those who are open to revelation. Ellis speculates that quantum events in the brain (directed by God) might be amplified to produce revelatory thoughts, images, and emotions. If it is supposed that God has adequate reason to restrict divine action to a combination of ordinary action (in and through natural processes) and revelation (such as the Resurrection of Christ) then the problem of evil does not take on the same dimensions as it does when it is assumed that God might freely intervene in any sort of process at any time."

Comment: Good try.

Bernd-Olaf Küppers, "Understanding Complexity"

"Epistemic reductionism leads to ontological reductionism in which 'life is nothing but a complex interplay of a large number of atoms and molecules.' Even consciousness must ultimately be reducible to physical laws. To counter this program, some biologists and philosophers of science appeal to 'emergence' and 'downward causation,' claiming that genuinely novel properties and processes arise in highly complex phenomena. According to this view, physics is a necessary part of the explanation but it cannot provide a sufficient explanation on its own. Küppers summarizes the claims of emergence and downward causation, respectively, as follows: '(1) The whole is more than the sum of its parts. (2) The whole determines the behavior of its parts.'

"Küppers concludes that 'both (emergence and downward causation) must be thought of as characteristics of self-organizing matter that appear at all levels when matter unfolds its complexity by organizing itself.'"

John Polkinghorne, "The Metaphysics of Divine Action"

"Polkinghorne prefers an approach based upon interpreting the unpredictabilities of chaotic dynamics (in accord with the strategy of critical realism) as indicating an ontological openness to the future whereby 'active information' becomes a model for human and divine agency. He interprets sensitivity to small triggers as indicators of the vulnerability of chaotic systems to environmental factors, with the consequence that such systems have to be discussed holistically. It is *not* supposed, however, that such triggers are the local mechanism by which agency is exercised."

Comment: This seems to be a change from his earlier views.

Nancey Murphy, "Divine Action in the Natural Order: Buridan's Ass and Schrödinger's Cat"

"Murphy argues that the problem of divine action will be solved by nothing less than a revised metaphysical theory of the nature of matter and of natural causes. Her proposal is that we view the causal powers of created entities as inherently incomplete. No event occurs without divine participation but, apart from creation *ex nihilo*, God never acts except by means of cooperation with created agents.

"She claims that [criteria, derived from both theology and science, which any satisfactory theory of divine action must meet] must allow for objectively special divine acts, yet not undercut our scientific picture of the law-like regularity of many natural processes.

"Murphy's proposal is that any adequate account of divine action must include a 'bottom-up' approach: if God is to be active in all events, then God must be involved in the most basic of natural events. Current science suggests that this most basic level is quantum phenomena. It is a bonus for theology that we find a measure of indeterminacy at this level, since it allows for an account of divine action wherein God has no need to overrule natural tendencies or processes. This cooperation rather than coercion is in keeping with God's pattern of respecting the integrity of other higher-level creatures, especially human creatures.

"One of these consequences is that the 'laws of nature' must be descriptive, rather than prescriptive; they represent our human perceptions of the regularity of God's action."

Comment: Murphy seems to agree with me about the laws of nature being human inventions. So, there are at least two of us.

Arthur Peacocke, "Chance and Law in Irreversible Thermodynamics, Theoretical Biology, and Theology"

"Peacocke sees chance as the means by which all possibilities for the organization of matter are explored in nature."

Arthur Peacocke, "God's Interaction with the World: The Implications of Deterministic 'Chaos' and of Interconnected and Interdependent Complexity"

"Peacocke concludes that, whatever is decided about those effects, the unpredictabilities for us of non-linear chaotic and dissipative systems do not, as such, help us in the problem of articulating more coherently and intelligibly how *God* interacts with the world, illuminating as they are concerning the flexibilities built into natural processes. The discussion is based in part on the assumption that God logically cannot know the future, since it does not exist *for* God to know."

Comment: This seems to conflict with the Augustinian notion of God being timeless. Furthermore, it assumes a fundamental arrow of time, which we have seen cannot be found in physics except as a definition.

VOL. 3. *EVOLUTION AND MOLECULAR BIOLOGY* (1998)

Ian G. Barbour, "Five Models of God and Evolution"

Barbour "outlines four philosophical issues which characterize the interpretation of evolution. *Self organization* is the expression of built-in potentialities and constraints in complex hierarchically-organized systems. This may help to account for the directionality of evolutionary history without denying the role of law and chance. *Indeterminacy* is a pervasive characteristic of the biological world. Unpredictability sometimes only reflects human ignorance, but in the interpretation of quantum theory, indeterminacy is a feature of the microscopic world and its effects can be amplified by non-linear biological systems. He also argues for *top-down causality* in which higher-level events impose boundary conditions on lower levels without violating lower-level laws and he places top-down causality within the broader framework of holism. He distinguishes between methodological, epistemological, and ontological reduction. *Communication of information* is another important concept in many fields of science, from the functioning of DNA to metabolic and immune systems and human language.

"According to Barbour, each of these has been used as a non-interventionist model of God's relation to the world in recent writings. If God is *the designer of a self-organizing process* as Paul Davies suggests, it would imply that God respects the world's integrity and human freedom. Theodicy is a more tractable problem if suffering and death are inescapable features of an evolutionary process for which God is not directly responsible. But do we end up with the absentee God of deism? The neo-Thomist view of God as *primary cause* working through secondary causes as defended by Bill Stoeger tries to escape this conclusion, but Barbour thinks it undermines human freedom. Alternatively, God as providential *determiner of indeterminacies* could actualize one of the potentialities present in a quantum probability distribution. Selection of one of the co-existing potentialities would communicate information without energy input, since the energy of the alternative outcomes is identical. Does God then control all quantum indeterminacies—or only some of them?"

Comment: An attempt to avoid deism.

Paul Davies, "Teleology without Teleology: Purpose through Emergent Complexity"

"Paul Davies offers us a modified version of the uniformitarian view of divine action. In selecting the laws of nature, God chooses specific laws which allow not only for chance events but also for the genuine emergence of complexity. He claims that the full gamut of natural complexity cannot be accounted for by neo-Darwinism, relativity, and quantum mechanics; one must also consider nature's inherent powers of self-organization based on, though not reducible to, these laws. Still the emergence of complexity does not require special interventionist divine action.

"God selects the laws of nature; being inherently statistical, they allow for chance events at the quantum or chaos levels as well as for human agency. God need not violate these laws in order to act, and there is room for human freedom and even for inanimate systems to explore novel pathways. He then argues that quasi-universal organizing principles will be found to describe self-organizing, complex systems. They will complement the laws of physics, but they would not be reducible to or derivable from physics, nor would they refer to a mystical or vitalistic addition to them.

"Chance in nature is God's bestowal of openness, freedom, and the natural capacity for creativity. The emergence of what he calls the 'order of complexity' is a genuine surprise, arising out of the 'order of simplicity' described by the laws of physics.... The acid test, according to Davies, is whether we are alone in the universe. If the general trend of matter toward mind and culture is written into the laws of nature, though its form depends on the details of evolution, we would expect that life abounds in the universe. This accounts for the importance of the SETI [Search for Extraterrestrial Intelligence] project."

GOD ACTING AGAINST GOD

The premise keepers seek a God who does not violate laws of nature. These acts might be in response to earnest prayers, or the need to fix up some sequence of events that has gone off course just because of the large amount of random, unpremeditated chance that evidently exists in our universe. As Polkinghorne has put it, if God worked against the laws of nature it would be God acting against God, the presumed author of those laws.[10] So it is not simply a matter of saying God is God—he can do anything he wants to do. Surely God could exempt himself from any law he writes. But then, if he does this on a regular basis, we humans should be expected to empirically detect such actions in ways that I discussed in *God: The Failed Hypothesis.* Whatever actions the premise keepers propose for God to take in the current world, they seriously attempt to make them consistent with the laws of nature—at least as we perceive them on the human scale. This is not a restriction on God; it is a restriction on the possible theories of God that certain theologians wish to consider. In this scheme, what may appear as a miracle is just an unusual event—not a violation of natural law.

Another restriction on theologians is that their theories of God must allow for human free will, which is fundamental to Christian teaching. This means that God's actions might be thwarted by human actions. Somehow theologians have to arrange it so that divine action is beyond the reach of human capability to undo.

ACTING IN PHYSICS

As discussed earlier, the possibility remains that the universe is deterministic, in which case we would have the Enlightenment deist god back again. For the rest of this chapter let us ignore that possibility and stick with the conventional interpretation of quantum mechanics as a statistical theory that only determines the behavior of ensembles of systems.

Does the uncertainty principle of statistical quantum mechanics open up a place for God to act, poking his finger in so that a particle goes where he wants rather than, as implied by quantum mechanics, some nonpredetermined place? Many premise keepers have suggested so, with William Pollard, a physicist turned Episcopalian priest, setting the agenda in 1958.[11] For a complete history including a good discussion of Pollard's views, see the book by Saunders.[12] I will focus on more current work.

Recall the example given in chapter 8 of a free electron (that is, an electron not bound in an atom) initially confined to a region the size of an atom. We saw that in six seconds it could be anywhere within a volume the size of Earth. In this case, God could direct the motion of that electron to where he wants it within the limits of the uncertainty principle. By limiting himself to placing the electron at a precise location within a volume the size of Earth in six seconds, humans would not be able to detect that fact.

But note that God is in fact still violating a law of physics when he steps in. That violation is simply not detectable to humans. So this proposal still breaks Polkinghorne's dictum. God is acting against himself. Also note that by limiting himself to placing the electron within a finite region of space, he is surrendering some of his omnipotence.

Every gram of matter contains a trillion-trillion electrons, protons, and neutrons. The visible universe contains 10^{79}. The universe beyond our horizon contains at least 10^{130} of these particles. Multiply these numbers by a billion to get the number of photons and neutrinos. This means that the deity has to somehow maintain control over countless events taking place at the submicroscopic level over extended periods of time.

The prospect of God micromanaging all these particles throughout the universe (and perhaps many other universes) does not appeal to many of today's theologians. They are looking for ways for God to act on the everyday scale of human experience where that action is meaningful to humanity. If God is to act in the universe, those actions must be amplified by some mechanism and, what is more, they must involve large-scale phenomena that are otherwise not predetermined.

BUTTERFLIES AND CHAOS

Polkinghorne and others have proposed that the amplification mechanism might be found in the so-called butterfly effect of chaos theory, which we discussed in chapter 10. In this scheme, working within the uncertainty principle, God changes the initial conditions of a chaotic system to affect the outcome. Recall that chaos theory is fundamentally deterministic and that the unpredictability associated with outcomes in chaotic systems can be traced to their extreme sensitivity to initial conditions that are often not known with sufficient accuracy. Not having human limitations, God can presumably set the initial conditions with unlimited precision and, knowing how to do Newtonian mechanics better than we do and presumably having the best computer in heaven at his disposal, he will obtain his desired outcome. However, suppose those initial conditions have quantum uncertainties. Then, as we noted above, God acting "within the uncertainty principle" would still violate natural law and simply keep that fact hidden from us.

Christian schoolmaster Timothy Sansbury has pointed to three other problems associated with this scenario for God's action.[13] First, a significant time delay is involved in the kinds of chaotic amplification systems we might consider. For example, it might take several days for the butterfly effect to change the weather. Chaos amplification certainly would not move fast enough to change the course of a tornado heading straight for your house in answer to your prayer, or end a storm endangering a ship at sea. Of course, God, knowing everything, might anticipate those prayers. But then he also would have foreknowledge of supposedly free will events, in which case they are not free.

Second, it is not clear that dramatic changes, such as bringing rain in response to farmers' prayers, can be effected in this manner.

Third, during the time that a chaotic system is working its way from initial conditions to final outcome, something might happen to change that course. This may not be a butterfly flapping its wings, but since we are assuming God gave humans free will, some human might take an action that God did not anticipate when he made his adjustment to the initials conditions. For example, that human might decide at the last moment to get into his carbon monoxide-emitting SUV and drive to Las Vegas, changing the chemical composition of the atmosphere just enough to thwart God's plan.

God could of course decide to limit human free will. However, this would violate one of the premises to be kept, namely, a high level of human free will. Furthermore, preventing human interference with his desires still would not guarantee sufficient time for his quantum diddling to produce his desired macroscopic outcome. And, as already noted, most human-scale systems are not chaotic and so chaos amplification would not apply to them.

Sansbury adds:

> Even if quantum mechanics does offer the space for ongoing divine action without any breakdown of natural law, it does not provide an answer to the underlying problem of how divine action can be responsive to indeterminate events. If the response must come after the event, quantum mechanics implies that divine responses will usually be delayed even if delay is inappropriate to the situation and sometimes will fail altogether. If the action is originated before the event, the implication is either that God knows the final states of future indeterminate events, which is presumed to be contrary to true indeterminacy, or that God acts on presumptions about indeterminate events and therefore can be wrong or thwarted by other indeterminate events. In either case, the problem of avoiding a God who tinkers or who controls from the past is not solved.[14]

In short, while quantum mechanics with chaotic amplification may provide a place for God to act to change a natural event, it will not always prove possible, rendering God as somewhat less than omnipotent. More important, these still involve violations of God's laws.

Notice that if God acted at the quantum level, even though he would not be able to affect the outcome of every event, he could for some. For example, he might have easily prevented the evolution of smallpox from cowpox and the AIDS virus from simian immunodeficiency disease. So the theodicy problem—the problem of evil—remains, even in the light of God's less-than-full omnipotence.[15]

From this physicist's point of view, it appears that direct involvement in quantum and chaotic processes does not provide a viable, effective process by which God can act without violating his own laws of nature. At best he can only hide those violations. Furthermore, they will not work in all situations. This does not rule out, however, the possibility of a deist god who created the universe and endowed upon it the ability to act creatively to carry out his plans.

NOTES

1. William G. Pollard, *Chance and Providence: God's Actions in a World Governed by Scientific Law* (London: Faber and Faber, 1958), p. 12.

2. Victor J. Stenger, *The Comprehensible Cosmos: Where Do the Laws of Physics Come From?* (Amherst, NY: Prometheus Books, 2007), pp. 312–19; see also "A Scenario for a Natural Origin of Our Universe," *Philo* 9, no. 2 (2006): 93–102.

3. See, for example, D. J. Donaldson, H. Tervahattu, A. F. Tuck, and V. Vaida, "Organic Aerosols and the Origin of Life: An Hypothesis," *Origins of Life and Evolution of the Biosphere* 34 (2004): 57–67.

4. Victor J. Stenger, "The Premise Keepers," *Free Inquiry* 23, no. 3 (Summer 2003); *Has Science Found God?* (Amherst, NY: Prometheus Books, 2003), ch. 11.

5. Robert John Russell, Nancey Murphy, Arthur Peacocke, and C. J. Isham, eds., *Quantum Cosmology and the Laws of Nature: Scientific Perspectives on Divine Action* (Vatican City: Vatican Observatory Publications, 1993); Robert John Russell, Nancey Murphy, and Arthur Peacocke, eds., *Chaos and Complexity: Scientific Perspectives on Divine Action* (Vatican City: Vatican Observatory Publications, 1996); Robert John Russell, William R. Stoeger, eds., *Evolutionary and Molecular Biology: Scientific Perspectives on Divine Action* (Vatican City: Vatican Observatory Publications, 1998); Robert John Russell, Nancey Murphy, Theo C. Meyering, and Michael A. Arbib, eds., *Neuroscience and the Person: Scientific Perspectives on Divine*

Action (Vatican City: Vatican Observatory Publications, 1999); Robert John Russell, Philip Clayton, Kirk Wegter-McNelly, and John Polkinghorne, eds., *Quantum Mechanics: Scientific Perspectives on Divine Action* (Vatican City: Vatican Observatory Publications, 2001). For information on the series and a good summary of each talk in the first three volumes, go to http://counterbalance.net/ctns-vo/index-frame.html (accessed May 1, 2008).

6. *Zygon* 41, no. 3 (September 2006): 501–778.

7. Nicholas Saunders, *Divine Action and Modern Science* (Cambridge: Cambridge University Press, 2002).

8. Wesley J. Wildman, "The Divine Action Project, 1988–2003," *Theology and Science* 2, no. 1 (2004): 31–75; "Further Reflections on the Divine Action Project," *Theology and Science* 3, no. 1 (2005): 71–83.

9. Diarmuid O'Murchu, *Quantum Theology: Spiritual Implications of the New Physics* (New York: Crossroad, 1997); John Polkinghorne, *Belief in God in the Age of Science* (New Haven, CT: Yale University Press, 1998); John Polkinghorne, *Quantum Physics and Theology: An Unexpected Kinship* (New Haven, CT: Yale University Press, 2007); Philip Clayton, *Mind and Emergence: From Quantum to Consciousness* (Oxford: Oxford University Press, 2004); Ted Peters and Nathan Hallanger, eds., *God's Action in Nature's World: Essays in Honour of Robert John Russell* (Williston, VT: Ashgate, 2006).

10. John Polkinghorne, "The Metaphysics of Divine Action," in *Chaos and Complexity: Scientific Perspectives on Divine Action*, pp. 147–56.

11. Pollard, *Chance and Providence.*

12. Saunders, *Divine Action and Modern Science.*

13. Timothy Sansbury, "The False Promise of Quantum Mechanics," *Zygon* 42, no. 1 (March 2007): 111–12.

14. Ibid., p. 112.

15. Thanks to Brent Meeker for pointing this out.

15

THE GOD WHO PLAYS DICE

I shall never believe that God plays dice with the world.
—Albert Einstein

GOD, CHANCE, AND PURPOSE

After carefully perusing much of the theological literature referenced in the previous chapter, I conclude that theologians have not solved the problem of divine action and they know it. However, an outsider, David J. Bartholomew, Emeritus Professor of Statistics at the London School of Economics and Political Science, has proposed a new direction to look.

In *God, Chance, and Purpose*, Bartholomew examines the role of chance in theology.[1] Theologians have generally viewed chance as anathema to religion. They see it as the enemy of purpose. On the other hand, today's scientists see chance in physical phenomena on all scales, from the cosmic to the biological to the subatomic.

Most of the matter of the universe moves about randomly. For example, consider the cosmic microwave background that pervades all of space. This is composed of very low-energy photons that outnumber the atoms in stars and planets by a factor of a billion. They constitute a gas in thermal equilibrium at 2.7° Celsius above absolute zero (2.7 Kelvin) that is distributed randomly to within one part in a hundred thousand. An even greater number of neutrinos are out there with similar lack of structure.

In the microworld, atomic transitions and nuclear decays occur by chance. In the world of living organisms, chance works with natural selection as the mechanism for biological evolution. Thirty years ago the Nobel Prize–winning biologist Jacques Monod wrote, "Chance alone is the source of every innovation, of all creation, in the biosphere. Pure chance, absolutely free but blind is at the very root of the stupendous edifice of evolution."[2] This remains the view of most scientists, but only a few such as Monod, Steven Weinberg, and Richard Dawkins have the courage to say so publicly. And they are castigated for it.

In the opposite corner, Calvinist theologian R. C. Sproul writes in *Not a Chance: The Myth of Chance in Modern Science*, "If chance exists in its frailest possible form, God is finished.... If chance exists in any size shape or form, God cannot exist."[3] Sproul cannot accept his own deduction and so concluded that the concept of chance as used in science is a myth. Either God exists and science is myth or science is right and God is a myth. Take your pick.

Other theists have attempted to downplay the role of chance in the world. In *A Case against Accident and Self-Organization*, lawyer Dean Overman claims to prove that accident and chance can have played no role in the formation of the universe and the origin of life, among other natural phenomena. He bases his conclusion on very dubious estimates of the probabilities for these events, without presenting the probabilities for these phenomena being supernatural. He makes the bizarre assertion that any chance event with a probability of less than one in 10^{50} is a "mathematical impossibility."[4] This is grossly incorrect, as several authors including philosopher Graham Oppy, Bartholomew, and I have shown.[5] Write down a sequence of fifty random numbers from 0 to 9. The a priori probability for the particular sequence of numbers is 1 in 10^{50}, hardly "mathematically impossible."

Bartholomew, a believer, stands at the center of the ring, claiming a place for chance in both science and theology. As we have seen, the premise keepers accept the role of chance in the world and look to it as a possible locus for divine action. Bartholomew views this as a too passive role for chance, which is seen as simply providing space for God to act. He does not regard as satisfactory either Sproul's proposal that chance is a myth or the premise-keeper proposal that God occasionally uses chance to act in the universe.

The statistics professor devotes most of his book to making the case for chance being central in God's providence, providing God with a "rich tapestry of opportunities and possibilities in the creative process." Chance is to be seen as "a real part of the creation and not the embarrassing illusion which much contemporary theology makes it out to be."[6]

Bartholomew bases his view on the way order arises by chance out of chaos in the macroworld of matter, just what we talked about in chapter 10. An example I like to give, also used by Bartholomew, is how the exponential decay "law" for radioactive nuclei follows from the fact that nuclear decays occur by chance. It can be shown by a simple mathematical derivation that if the probability for a nucleus decaying in a certain time interval is the same for all time intervals, as expected by chance, then the number of nuclei remaining from an original sample will fall off exponentially with time. This is a "law of physics," obeyed with great accuracy in many experiments; yet it follows from pure chance.

Bartholomew gives other examples of laws that follow from chance, such as the normal (Gaussian) and binomial frequency distributions. He has evidently not studied cellular automata, which we learned in chapter 10 provide remarkable examples of complexity arising from simplicity. However, Bartholomew does mention similar results obtained in the study of networks.[7]

While he does not use the term, Bartholomew's chance provides a mechanism for what is called "emergence," as also discussed in chapter 10.

Bartholomew illustrates how his view of a "creative" role for chance operates in evolution. He sees God not stepping in to guide each mutation along an evolutionary path, but allowing the creative processes of pure or "blind" chance, which God after all created, to operate.

The author must then contend with the problem of the human species evolving as part of God's plan, one of the dogmas of Christianity as well as other faiths. Bartholomew relies on a controversial and unconvincing proposal by paleontologist Simon Conway Morris that evolution will naturally converge on intelligent beings, "inevitable humans," regardless of the path taken in the space of all possibilities.[8]

CHANCE AND EVOLUTION

With the demise of the clockwork universe in which everything, including evolutionary paths, was predetermined, theologians had to grope for a place for God to act in the course of evolution. Since we still do not understand how life originally came about on Earth, anyone is free to propose that God initiated the process. However, even if God did create life, he could not simply then turn it over to natural selection the way the Enlightenment deist god turned over physics to Newtonian determinism.

Consider the matter of the development of the human species. Evolution tells us that we are the result of the natural selection out of an enormous number of random mutations that have occurred since life began on Earth four billion years ago. If, as Christianity and other religions teach, God created the universe with a special place and plan for humanity, then he would have had to step in countless times along they way—every time there was a mutation on the path to *Homo sapiens*, to make sure we evolved. Such actions may be indistinguishable from normal evolution but would constitute a form of intelligent design inconsistent with the more parsimonious evolutionary principle that random mutations are sufficient to provide the genetic changes that are needed for natural selection to operate. Furthermore, as with divine action in physics, each time God steps in he is violating his own laws of nature—the laws of evolution. That is, he has the same problem we saw in the last chapter, where even quantum uncertainty and chaos are insufficient for God to act in the physical world without violating what are, presumably, his own laws.

The famed paleontologist Stephen Jay Gould argued that evolution shows no evidence of any purposeful movement toward increasing complexity and sophistication. He inferred this from observations of data from the Burgess Shale that he analyzed in his sensational 1989 book, *Wonderful Life*.[9] In his familiar metaphor, Gould talked about rewinding the tape of evolution and running it over and over again. He claimed history would never end up at the same place twice, implying that humanity is a pure accident.

Conway Morris did much excellent research on the Burgess Shale.[10] Gould was effusive in his praise of Conway Morris's work and recommended him for a prestigious award. Conway Morris, however, has argued against Gould's metaphor with what only can be called religious fervor.

Conway Morris claims that evolution demonstrates what he has termed "convergence," where independent evolutionary paths often come up with the same solution to problems. This has led him to infer that the evolution of some form of intelligent life was inevitable in the course of evolution, specifically humanity if you take literally the subtitle of his 2003 book, *Life's Solution: Inevitable Humans in a Lonely Universe.*[11]

Bartholomew throws his chips in with Conway Morris, noting that the paleontologist has amassed an "extensive catalogue of examples" of convergence.[12] Bartholomew sees Darwinian evolution based just on chance and natural selection, being God's way of creating humanity with no other divine action necessary. Fine, but isn't this yet again the deist god rather than the Christian God?

While Conway Morris does indeed give many examples where different evolutionary paths seem to lead to similar solutions, such as the camera eye, he has not demonstrated their inevitability. As philosopher of science Elliot Sober points out in his *New York Times* review of *Life's Solution*, "You can't show that an event was inevitable or highly probable just by pointing out that it has happened many times. To estimate the probability of the camera eye's evolving, you need to know how many times it evolved and how many times it did not. Conway Morris never describes how often convergences failed to occur."[13] For example, he might have counted the ten or so different kinds of eye as cases of nonconvergence.

Sober also points out that probabilities multiply so that even if you have a long sequence of evolutionary transitions, each with a high probability, the probability for the sequence can still be very low. Conway Morris has committed the same fallacy that we find with creationists, who estimate very low probabilities for life having a purely natural source without providing comparative probabilities for that source being supernatural.[14]

If life is converging inevitably toward high intelligence, how is it that of the millions of species on Earth we have only one, *Homo sapiens*, with that ability? Look at how far ahead we are than other animals in intelligence. Indeed, since most life forms are microbial, intelligence would not seem to be very high at all on the universe's agenda.

THE EMERGENCE OF SELF-ORGANIZATION

Nevertheless, as I have noted many times now, simplicity begets complexity in the natural world. What both Bartholomew and Conway Morris may be seeing in living organisms and their fossils is the more general process by which nature builds different levels of complex systems with new properties and principles at each level by means of the process termed *emergence*, which was discussed in detail in chapter 10. Biology is one such level, and the principles of evolution are surely emergent.

There is little argument among scientists and theologians that the main process of emergence is "bottom-up" causality, that the systems and all levels are still governed by the basic laws of particle mechanics. More controversially, the notion that some additional form of "top-down" causality in which a system with a high level of complexity is able to act downward and affect lower levels in a causative fashion is being promoted by those who think this will show that we humans are more than "just particles."

Evolution is cited as having this property, but I have not seen any example where top-down causality in biology is not as trivial as a rotating wheel causing its constituent particles to move in a circle. I have cited physicist Paul Davies as a representative of the materialist emergence school. In his earlier writings at least, he argued that the emerged principle of self-organization was "a fundamental law in its own right" that could not be derived from the fundamental laws of mechanics.[15] As we have seen, however, the fact that most higher-level principles cannot be strictly derived from lower-level ones does not necessarily mean they are still not the result of actions at the lower level. Computer simulations seem to bear this out.

Both Davies and Stuart Kauffman see self-organization as a creative force operating in the universe driving life higher and higher. Although in their previous writings they were probably unfamiliar with Bartholomew's work, I imagine they would agree with him on the large role played by chance.

Kauffman sees the creative force acting without any supernatural element. As we saw in chapter 14, where we reported on Davies's contribution to the Vatican conference, he also sees no need to bring God into the picture.

But no God is not an option open to theologians, and Bartholomew is a member of the holistic emergence camp, who see a role for the divine.

EMERGENCE AND GOD

Philip Clayton has speculated about the possible places where emergence may have religious significance.[16] He suggests that a deity may be seen as another level of emergence, presumably the ultimate level. This idea is similar to the *Omega Point God* proposed in 1955 by the Jesuit priest and paleontologist Pierre Teilhard de Chardin.[17] Teilhard saw evolution beginning at the *Alpha Point*, a place of infinite disorder, and proceeding by increasing complexification to humanity and onward to the Omega Point that is Christ. Perhaps he should be given credit as one of the earlier discoverers of emergence.

However, few theologians are interested in a God who is far in the future. Physicist Frank Tipler has estimated the time we need to wait to reach the Omega Point as a billion-billion (10^{18}) years.[18] Clayton suggests a gradual process, an increasing "deification" of the universe over time.[19] Of course, the Omega Point Christ is not much like the Christ of the Bible and Clayton conjures up a sort of theological dualism with the mind emerging physically toward a greater knowledge of the metaphysical God who always was. Assuming the nontrivial top-down causality (holistic emergence), God-to-mind divine action can occur without the need to manipulate particles or to break physical laws.

We might even apply this idea to all complex systems in the universe. Again with top-down causality, God is able to intervene at a high level and affect events at lower levels without the problems we found with the direct intervention at the atomic level and the need to use quantum uncertainty or chaotic amplification.

But this is all highly speculative and dependent upon nontrivial top-down causality, for which there is no evidence.

THE NEW DEIST GOD

Still, emergence opens up another possible route for divine action in the world that is suggested by the work of Bartholomew discussed above as well as by several theologians. Without tinkering in the process, god (the unacknowledged deist god) bestowed a rich creativity on the cosmos at the creation because his laws have the ability to facilitate the evolution of matter and energy along pathways of ever-increasing complexity.

God may not intervene but still affect events. As theologian Thomas Tracy puts it:

> The *given* potentialities of nature are given *by* God, not *to* God. God acts most fundamentally by establishing and sustaining the structures of nature, and only secondarily by redirecting events within those structures, This divine creative activity sets the direction of cosmic history and so is the primary mode of God's providential governance of the world.[20]

Tracy obviously does not want to identify this as the deist god, so he allows for God to "secondarily" redirect events. But how can God do this without confronting all the problems of divine action we have already covered in detail?

While this may explain how a deist god can produce complexity without tinkering, it is hard to see how a general principle of increasing complexity could be preordained to lead to the evolution of any specific species such as humanity.

Some theologians say it is not important whether humans evolved or not. For his own reasons, the deity set things up the way he did, with countless paths to some final end that need not include humanity.[21] That's possible, but such a god must then be reconciled with the traditional Christian teaching and widespread human belief that we are special and central to the divine plan for the universe. As theologian Keith Ward explains:

> The Christian faith is committed to the belief that there is a God who acts in history. In the Old Testament God liberates the Hebrews from Egypt by acts of might and terrible power. He guides them through the wilderness in a pillar of cloud and fire. He gives them victory over their enemies by sending plagues, earthquakes and inducing supernatural panic. He allows his people to be taken into exile because of their sins, and helps them return to the promised land at the time he has foretold. The core of the New Testament is a record of the acts of God in the life, death and resurrection of Jesus. It is by the power of God that Jesus heals, exorcises demons, and is raised from the dead in vindication of his prophetic ministry. In the daily life of churches, Christians pray, by their Lord's command, for God's help and guidance; and believe that God can and will act to heal, convert and empower human lives.[22]

One possible means of reconciliation might be to suggest that the universe is so vast that humans would evolve somewhere. Still, you would think that a creator with infinite intelligence would come up with a simpler solution than creating trillions upon trillions of planets so that on one he could send down his son to redeem the sins of one particular bipedal hominoid out of millions of other species of living things that accidentally evolved.

Nevertheless, this may be the only way divine purpose can be served. It leaves ample room for a great amount of creativity that is possible given the way simple systems are able to evolve into more-complex forms naturally and without any outside help. Humans are not part of that picture except as a species that just happened to evolve by accident. As physicist and theologian Ian Barbour has put it, "There can be purpose without an exact predetermined plan."[23] Theologian John Haught similarly remarks, "A God whose very essence is to be the world's open future is not a planner or a designer but an infinitely liberating source of new possibilities and new life."[24] This includes, by the way, the possibility of destroying all human life on Earth by nuclear war or global warming.

Indeed, so many Christian theologians seem to be converging on this view of god that they may be a developing consensus. But few if any have owned up to the fact that they have abandoned the Christian God in favor of a new deist god, where

1. God does not directly act in the universe.
2. *Homo sapiens* is an accident.
3. Much of what happens in the universe is not according to specific plan.

The new deist god creates the universe but includes in it a huge element of chance. Of course apologists will work very hard to reconcile this new god with Christianity, but that will take an awful lot of angels and an awful lot of pinheads. It is hard to see how they will make this god into one who still should be worshipped and prayed to.

The new deist god is precisely the one Einstein objected to when he said (on numerous occasions), "I shall never believe that God plays dice with the world."

The premise keepers may feel satisfied that at least their new god created the universe with some plan and that we are now living out that plan.

But I have bad news for them. Modern physics and cosmology imply that all the creator did when he made the universe, if he existed at all, was make a single toss of the dice.

NOTES

1. David J. Bartholomew, *God, Chance, and Purpose: Can God Have It Both Ways?* (Cambridge: Cambridge University Press, 2008).

2. Jacques Monod, *Chance and Necessity*, trans. Austryn Wainhouse (New York: Collins, 1972), p. 110.

3. R. C. Sproul, *Not a Chance: The Myth of Chance in Modern Science* (Dartmouth, MS: Baker Books, 1994).

4. Dean L. Overman, *A Case against Accident and Self-Organization* (New York: Rowman & Littlefield, 1997), p. 1.

5. Graham Oppy, "Review of Dean L. Overman," in *A Case against Accident and Self Organization*, online on the *Secular Web* (1999) http://www.infidels.org/library/modern/graham_oppy/overman.html (accessed May 14, 2008); Bartholomew, God, Chance, and Purpose, p. 2; Victor J. Stenger, *Has Science Found God? The Latest Results in the Search for Purpose in the Universe* (Amherst, NY: Prometheus Books, 2003), p. 322; Bartholomew, *God, Chance, and Purpose*, p. 2.

6. Bartholomew, *God, Chance, and Purpose*, p. 15.

7. Ibid., pp. 40–42.

8. Simon Conway Morris, *Life's Solution: Inevitable Humans in a Lonely Universe* (Cambridge: Cambridge University Press, 2003).

9. Stephen Jay Gould, *Wonderful Life: The Burgess Shale and the Nature of History* (New York: Norton, 1989).

10. Simon Conway Morris, *The Crucible of Creation: The Burgess Shale and the Rise of Animals* (Oxford: Oxford University Press, 1998).

11. Conway Morris, *Life's Solution.*

12. Bartholomew, *God, Chance, and Purpose*, p. 188.

13. Elliot Sober, "It Had to Happen," *New York Times*, November 30, 2003.

14. See my discussion in Victor J. Stenger, *God: The Failed Hypothesis—How Science Shows That God Does No Exist* (Amherst, NY: Prometheus Books, 2007), p. 138.

15. Paul Davies, *The Cosmic Blueprint: New Discoveries in Nature's Creative Ability to Order the Universe* (New York: Simon & Schuster, 1988); paperback ed. (Radnor, PA: Templeton Foundation Press, 2004), p. 144.

16. Philip Clayton, *Mind and Emergence: From Quantum to Consciousness* (Oxford: Oxford University Press, 2004).

17. Pierre Teilhard de Chardin, *The Phenomenon of Man*, English trans. Bernard Wall (London: Wm. Collins Sons; New York: Harper & Row, 1959). Originally published in French as *Le Phénomene Humain* (Paris: Editions du Seul, 1955).

18. Frank J. Tipler, *The Physics of Immortality: Modern Cosmology and the Resurrection of the Dead* (New York: Doubleday, 1994).

19. Clayton, *Mind and Emergence*, p. 167.

20. Thomas F. Tracy, "Divine Action and Quantum Theory," *Zygon* 35, no. 4 (December 2000): 891–900.

21. Kenneth R. Miller, *Finding Darwin's God: A Scientist's Search for a Common Ground between God and Evolution* (New York: HarperCollins, 1999), p. 213; John F. Haught, *God after Darwin* (Boulder, CO: Westview Press, 2000), p. 120.

22. Keith Ward, *Divine Action* (London: Collins Religious Publishing, 1990), p. 1.

23. Ian G. Barbour, *Religion and Science* (New York: HarperCollins, 1997), p. 216.

24. Haught, *God after Darwin*, p. 120.

16

NOTHINGISM

On the far side of the Big Bang is a mystery so profound that physicists lack the words to even think about it. Those willing to go out on a limb guess that whatever might have been before the Big Bang was, like a vacuum, unstable. Just as there is a tiny chance that virtual particles will pop into existence in the midst of sub-atomic space, so there may have been a tiny chance that nothingness would suddenly be convulsed by the presence of a something.

—Robert Crease and Charles Mann[1]

In this and in previous books I have tried to show that both observational data and mathematical theory demonstrate beyond a reasonable doubt that no spirit world exists. The universe is truly comprehensible as a purely material system. We can fit all observations to a model of elementary particles (or perhaps strings or other forms of basic objects) that move around in an empty void—just as the Greek atomists' conjectures from thousands of years ago that were almost lost to history.

Atomism resurfaced in Europe in the Middle Ages, where it formed part of the basis of Newtonian mechanics. The atomic model was further buttressed by the physics and chemistry of the nineteenth century, finally and definitively being verified by the developments of the twentieth century, as I have reviewed in this book. The conclusion that everything is matter and nothing more is also confirmed by the developments in other fields such as neuroscience that I have not covered.

THE IMPOSSIBLE GOD

In *God: The Failed Hypothesis* I considered a God with the attributes of the traditional deity of the three great monotheisms, Judaism, Christianity, and Islam. This God created the universe with a special place for humankind. He is perfectly loving, moral, all powerful, and all seeing. I called him the "3O God": Omnibenevolent, Omnipotent, Omniscient. This God takes an active role in worldly events, stepping in often to change their course and to respond to human entreaties. As such, he performs miracles in which the laws of nature are violated. He also listens to every human thought, answering prayers when he wishes.

As I claimed in that book, this God can be shown beyond a reasonable doubt not to exist. I will not run through all the arguments but mention those that I regard as the most powerful. I begin with two logical arguments.

First is what theologians have called the *theodicy problem*: How can a perfectly loving, moral, all-powerful, and all-seeing deity allow evil such as unnecessary suffering to exist in the world?[2] Theologians have struggled with the question for centuries and have never come up with a satisfactory answer. The best that they can say is this is a "mystery" and that God must have some purpose in allowing evil to exist. The fact is, evil exists in the world and the 3O God has the power to prevent it. Thus, either such a God (1) lacks the power to prevent evil, (2) lacks the knowledge to know everywhere evil occurs, (3) is neither good nor loving so he purposefully allows evil in the world, or (4) does not exist.

A second compelling argument against the existence of the 3O God is called the *argument from nonbelief*.[3] A related form is called the *hiddenness argument*.[4] Many people lack belief in God, although they are perfectly open to the possibility of his existence and seek only some evidence, some sign. If God exists, he deliberately hides from these people, not providing them with any sign. Thus, either such a God does not exist or he is an immoral God who refuses to reveal himself to those who honestly seek to know him. Indeed, this is a God that many people worship, in particular, evangelical Christians and Muslims. He is not an impossible God, just a wicked one.

Many additional logical arguments for the impossibility of God can be found in the anthology edited by Michael Martin and Ricky Monnier.[5] The same editors have also produced a volume of improbability arguments.[6]

My main arguments against the existence of the Judeo-Christian-Islamic God are scientific ones. A God who plays such an important role in the universe as the God of the great monotheisms should leave observable, physical evidence for his existence. As we have seen, he cannot deliberately hide from us and remain a moral god. We should see evidence for God in the cosmos, in life on Earth, and in human activities. However, using our own senses and the scientific instruments we have developed to aid those senses, we find no evidence for God or any form of supreme spirit.

You often hear theists and even some reputable scientists say, "Absence of evidence is not evidence of absence." I dispute this. Absence of evidence can be evidence of absence beyond a reasonable doubt when the evidence should be there and is not found. For example, no one has seen elephants in Rocky Mountain National Park. But surely, if elephants did roam the park we should have evidence for them: footprints, droppings, smashed grass. While a remote possibility exists that they have remained hidden all this time, we can conclude beyond a reasonable doubt that Rocky Mountain National Park is not inhabited by any elephants.

In this manner, the absence of evidence for the Judeo-Christian-Islamic God where there should be clear evidence allows us to conclude, again beyond a reasonable doubt, that such a God does not exist.

The arguments I gave in *God: The Failed Hypothesis* were all based on the expectation that God's actions in the world, including the creation itself, should be detectable. Of course, an all-powerful God is free to do whatever he wants, but it is inconsistent with his omnipotence for God to violate presumably perfect laws that he instituted in the first place.

Furthermore, such a God would be much easier to detect by our direct observation of natural law violations. We see no such violations in cosmology, physics, biology, or other sciences. We can safely conclude that an omnipotent God who takes direct divine action in the world without violating natural law can be ruled out.

THE BATTLE BETWEEN SCIENCE AND SUPERSTITION

We began this book by looking at the current beliefs of Americans. We saw that while the nation is unusually religious compared to other wealthy countries, Protestants no longer constitute the majority. Furthermore, although 80 percent of Americans profess to be Christians, 44 percent do not believe in a God who plays a vital role both in the universe or in their personal lives. Their beliefs match more closely the doctrine of deism than theism.

We also found that an increasing percentage of Americans are abandoning traditional religion altogether, although they are not quite ready to give up on the idea of a spirit world beyond matter. Popular films and books over the past generation have promoted the notion that modern physics and particularly quantum mechanics have revealed a connection between human consciousness and reality that is purported to provide a scientific basis for a spiritual component to the universe. The message of self-help gurus who assemble an attractive package of Eastern mysticism mixed with modern physics is "You can make your own reality." Evidence to support this astounding notion is gleaned from reports of paranormal phenomena and the anecdotal success stories of various forms of alternative medicine. However, none of these claims stand up under critical scrutiny of the same kind that is applied in science to any proposed extraordinary phenomenon.

Unfortunately, public understanding of science and the scientific method (as well as many other important disciplines such as history and philosophy) is so inadequate that many people are easy prey to the charlatans who promise simple solutions to difficult and often unsolvable problems.

In *The Republican War on Science*, journalist Chris Mooney documents how the administration of George W. Bush and the anti-intellectualism of America's religious right systematically undermined science in almost every area of social importance from global warming to safe sex.[7] While many scientists have spoken out against the misuse of science, the majority have chosen not to get involved in the fray. I plead with my scientist colleagues to take a more active role in what fundamentally continues the ancient battle between science and superstition. Their noninvolvement may be the easy way out, but all it does is encourage irrational thinking that surely is not of benefit to society.

SOMETHINGISM

So far in investigating the possibilities for "something out there" beyond the material world, I have tried to stick to the empirical evidence. Where I have used theoretical ideas, all have been based on theories such as relativity, quantum mechanics, and the standard model of particles and forces that are well confirmed based on their ability to describe and predict observations with great precision. These have led me to the conclusion that if there were something out there that had any significant influence on the operation of the universe and the behavior of its inhabitants, science would by now have the data to prove it beyond a reasonable doubt. Since substantial divine intervention is fundamental to Judeo-Christian-Islamic belief, the God of these religions would seem to be ruled out by the data.

We have studied the attempts by Christian theologians to use quantum mechanics and chaos theory to provide a place for a deity's intervention in the world without violating his own laws of physics. We found that this was not viable. Even if God were to utilize these phenomena in ways that were undetectable to humans, he would still be violating the laws of physics in the process. Furthermore, macrosystems are not all chaotic and those that are involve time delays, so this option is not always available. To have full control over all events God would have to manage the motion of every fundamental particle in the universe in a nanosecond-by-nanosecond basis. I suppose, being omnipotent, he could do that. But I get the impression in my reading that most theologians would not be happy with that solution. Furthermore, such micromanagement would still not guarantee a predetermined outcome on the macroscale.

This leaves as the only possible moral God a deist god who creates the universe and then lets it go on its own way. Part of that creation leaves many events to chance so that much of what happens in the universe, including the evolution of biological species, is beyond this god's direct control. Humanity is then an accident, not a special creation. Many Christian theologians propose such a supreme being, but despite their valiant attempts they have so far failed to turn it into the Christian God. Deism and Christianity have always been totally incompatible—as incompatible as science and Christianity.

As a spiritual alternative to any god or God we have investigated the notion that the universe is permeated by an irreducible cosmic consciousness that includes the human mind. We found that the claim that quantum mechanics shows a connection between human consciousness and reality is a misunderstanding or deliberate misrepresentation of what quantum mechanics really says.

What's more, no empirical evidence can be found that the human mind possesses any of the special powers such as ESP or mind-over-matter that we would expect if some immaterial, supernatural component were involved.

For most people, this will not be a satisfactory end to the story. Even if I have shown that most common concepts of God and spirit are invalid, questions still remain. I still need to say something about: "Where did it all come from?" "Why is there something rather than nothing?" "Where did the laws of physics come from?"

QUANTUM TUNNELING

Physicists and cosmologists are often criticized for speculating about domains where they have no direct empirical data and, worse, where no measurements can physically be performed.[8] These critics are simply ignorant of science. This procedure is quite common and, furthermore, has an excellent track record of success.

The basic process that both illustrates this point and happens to also provide a plausible mechanism for the universe's natural origin is called *quantum tunneling*. In classical mechanics a body of a certain kinetic energy cannot surmount a barrier where the potential energy is higher than that kinetic energy. In quantum mechanics it is possible for that body to tunnel through the barrier.

For example, a dog can be confined to a yard if the fence is high enough. In principle he can quantum-tunnel through the fence, but calculation gives a very low probability for this happening on the macroscale. Nevertheless, quantum tunneling can and does happen on the microscale.

In figure 16.1 a particle of fixed energy and momentum is coming in from the left and is represented by a sinusoidal wave function. Mathematically, both energy and momentum are real numbers consistent with the fact that they are measurable, and in physics all measurements by convention are represented by real numbers. Inside the barrier, however, the momentum of the particle is an *imaginary number*, that is, has a negative square root and so is "unphysical," meaning unmeasurable.

The Schrödinger equation, which is used in elementary quantum mechanics to calculate wave functions, does not care if numbers are real or imaginary. In fact, the wave function itself is in general a *complex number* that has a real and an imaginary part (recall figure 12.3, p. 186). Inside the barrier the wave function is an exponential function indicating that the particle has some nonzero probability for being in that region. However, as explained above, that particle is "unphysical" since it has imaginary momentum. Once the exponential wave function reaches the edge of the barrier, the particle reappears with a measurable, real momentum. In the figure we see the particle emerging from the barrier is represented again by a sinusoid. The particle has the same energy as it did going in. However, the magnitude of the emerging wave function is lower. Squaring the ratio

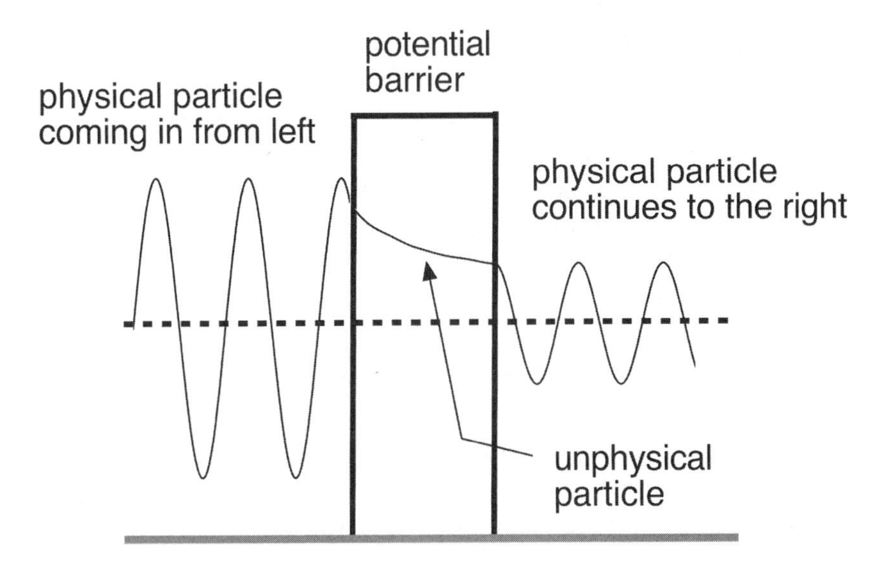

Fig. 16.1. The sine wave on the left represents a particle incoming from the left with energy given by the dashed line, which is less than the height of the barrier. The exponential curve inside the barrier represents a particle with an unphysical, imaginary momentum. It leaks through the barrier and a physical particle appears to the right. The ratio of the amplitudes of the outgoing and the incoming wave functions gives the probability that the particle will penetrate the barrier.

of the magnitudes of the two sine waves will give you the probability that the particle will tunnel through or be reflected back to the left.

The phenomenon of quantum tunneling has been well established for generations. It was first used by physicist George Gamow in 1928 to explain the *alpha-decay* of atomic nuclei, which it did successfully over a range of many orders of magnitude in decay half-lives. Today, it makes possible one of our most powerful modern devices, the *scanning tunneling microscope*, which can provide pictures of individual atoms.

THE WAVE FUNCTION OF THE UNIVERSE

In 1983 James Hartle and Stephen Hawking proposed what they called the "no boundary" model of the origin of the universe.[9] Following a review by David Atkatz,[10] I have worked a version of this model out with complete mathematical rigor, although I managed, with some simplifications, to do this at about the level of a senior physics or mathematics major at an American university. The details can be found in my book *The Comprehensible Cosmos*[11] and in my article in the philosophical journal *Philo*, available online.[12] Let me just summarize the procedure, which I hope will sufficiently convince those theists who have expressed skepticism and even ridiculed the notion.[13] Hartle, Hawking, and I are not just waving our arms but have a fully developed mathematical and physical proposal for how the universe can have come about naturally.

Start with the conventional equation derived from general relativity called the *Friedmann equation*, which describes the evolution of a spherically symmetric universe. Assume the universe is empty, that is, has no matter, but still contains the energy stored in the curvature of space that Einstein associated with what he called the *cosmological constant*. Apply the standard "quantization" technique used in quantum mechanics to go from a classical equation to a quantum one. The result is an equation that allows you to calculate a wave function that describes the state of the universe. Since the universe is empty and spherically symmetrical, the only variable is its radius.

This equation is a simplified version of what is called the *Wheeler-DeWitt equation*. Its form is mathematically identical to the nonrelativistic, time-independent, one-dimensional Schrödinger equation for a particle in a potential field familiar from elementary quantum theory. The particle has half the Planck mass,[14] zero total energy, and a specific potential energy that is defined in the book and the article. This does not mean the universe is a particle of half the Planck mass, just that its wave function is mathematically equivalent to that of this particle.

Hartle and Hawking, and others who have played this game, call this the *wave function of the universe.* Their particular solution is shown in figure 16.2. When the radius of the universe is greater than a certain fixed value, the wave function oscillates like a real particle, just like the particle outside the barrier in figure 16.1. This gives the exponential inflation that, according to modern cosmology, takes place during the first tiny fraction of a second before the conventional big bang. When the radius is less than this value, the wave function of the universe is in a nonphysical region analogous to the region inside the barrier in figure 16.1. Only now, time is represented by an imaginary number.

The Hartle-Hawking model of the natural origin of our universe describes a larger universe that has no beginning or end of time. This is consistent with our discussion of time in chapter 5. Out of the limitless past in the time before our big bang, assuming the arrow of time for our universe, this prior universe deflates to the point where it becomes unphysical and time is imaginary. Its wave function then tunnels through the unphysical region and our universe appears on the other side.

Note, however, the entropy in the "prior" universe increases in the opposite direction to ours; thus, the arrow of time in that universe points the opposite way. So it is really a mirror universe to ours, though not an exact image because of randomness. Both universes can be seen as emerging from the same chaos, one expanding in one time direction and the other expanding in the opposite direction. They both have a beginning after all! But it is still a causeless beginning.

Of course, talking about time having two directions throws most theologians into a tizzy. All theological discussions about creation assume an absolute direction of time and causality that is fundamentally wrong.

Alexander Vilenkin has proposed an alternate scenario in which no prior universe exists, but our universe simply tunnels one way out of chaos.[15] The same mathematical procedure applies in this case.

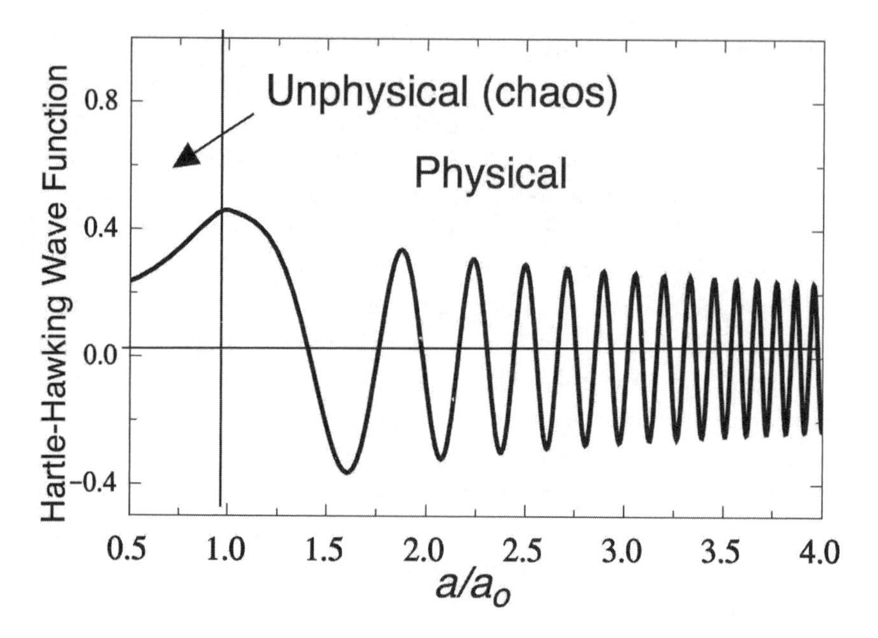

Fig. 16.2. The Hartle-Hawking wave function of the universe. The horizontal axis is crudely the radius of the universe a divided by a critical radius a_0. Above the critical radius the universe is physical. Below ($a/a_0 < 1$) the universe is unphysical. The unphysical region has no structure or information and so is interpreted as "chaos," or "nothing."

CHAOS AT THE ORIGIN

Let's talk about this "unphysical region." Inside this region no measurements can be made. Thus we have zero information about what is inside, which means the inside has maximum entropy (recall our discussion of Shannon information in chapter 5). Thus our universe (and its sister, if it has one) begins in a state of maximum disorder, or total chaos. This chaos is not to be confused with the "deterministic chaos" discussed in chapter 10. This is real, indeterministic chaos.

While the specific Hartle-Hawking and Vilenkin scenarios each have the universe emerging from total chaos, this notion is more generally a part of most cosmological origin scenarios where the universe exponentially explodes from the tiniest possible volume. This is simply what you get when you extrapolate the big bang back as far as you can go, the Planck time. Note that this is not the point "singularity" of infinite density inferred from general relativity and still mistakenly referred to by many theologians and others, as we discussed in chapter 5. Rather, a combination of general relativity and quantum mechanics leads us to conclude that this volume was of Planck dimensions and finite energy. While the two theories have not yet been reconciled in a single unified theory, there is no reason to believe that this conclusion will change with that theory. Such a volume would be too small to allow any physical quantities to be operationally defined (that is, be measurable even in principle) so we can conclude it is a region of maximum entropy.

Here's another way to look at it. Suppose the universe is a sphere, which we would expect from rotational symmetry. The maximum entropy of a sphere is that of a black hole of the same radius. A sphere with a radius equal to the Planck length is a black hole, so its entropy is maximum. Thus, the universe starts out as a black hole with maximum entropy.

Now, it is often asserted that the universe must have begun with a high degree of order since the second law of thermodynamics would require its entropy or disorder to increase with time. So, how could it have begun with maximum entropy? This apparent paradox is accounted for by the expansion of the universe. In any scenario where the universe expands from Planck dimensions the entropy of the universe will increase with time, consistent with the second law, simply because the entropy of any isolated expanding volume increases as the volume increases. It can be shown that the entropy increase for an expanding relativistic gas is proportional to the radius. On the other hand, as noted above, the maximum entropy of any volume is that of a black hole of the same volume. The entropy of a black hole is proportional to its surface area, so it increases as the square of its radius.[16]

As illustrated in figure 16.3, the entropy of the universe increases with radius, consistent with the second law but not as fast as the maximum entropy. This leaves an increasing entropy gap in which orderly structures form. In that case, entropy decreases locally as the rest of the universe gains entropy.

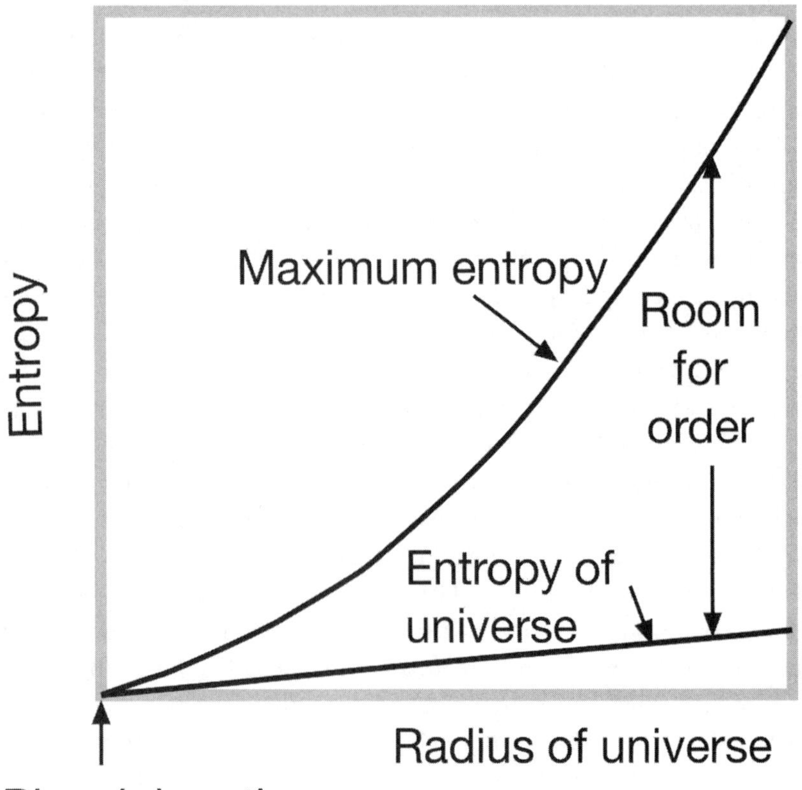

Fig. 16.3. At its origin the universe is as small as it could be with a radius equal to the Planck length, making it a black hole with maximum entropy. After the black hole disintegrated, the universe became an expanding gas whose entropy continues to increase. However, the maximum possible entropy increases faster so that there is increasing room for order to form locally.

OUT OF NOTHING

If we seek out the origin of the word *chaos* we find that it comes from the Greek for "the primal emptiness," or "void." To be "something" requires some structure. Since the unphysical region lacks everything, including structure, it can be identified as "nothing." Now, if you are what in the cyberworld is called a "logic chopper" you will try to argue that having no structure is still a property and anything with a property can't be nothing. Don't bother e-mailing me with such irrelevant wordplay. The unphysical region is as nothing as anything can be.

Now, a common question theists ask of atheists is "Why is there something rather than nothing?" This was posed in 1714 by the philosopher Gottfried Wilhelm Leibniz (and perhaps earlier) and is referred to as the *primordial existential question.*[17] In his 2004 book philosopher Bede Rundel concluded that there has to be something.[18] Philosopher Adolf Grünbaum has argued forcefully that the primordial existential question is ill-conceived because it assumes that the universe would be nothing in the absence of an overriding cause or reason.[19]

Let me give a complementary physics argument for why we should expect something rather than nothing, why something is the more natural state of affairs than nothing and so we would need God to maintain a state of nothingness rather than somethingness. In nature we find that in the absence of an external source of energy (heat), systems of higher symmetry tend spontaneously—without cause—to make phase transitions to states of lower symmetry where the energy is generally lower. For example, highly symmetric water vapor condenses into less symmetric liquid water, which then freezes into even less symmetric ice. As Nobel Prize–winning physicist Frank Wilczek noted in 1980, since nothing is more symmetric than nothing, you would expect nothing to be unstable.[20]

In the Hartle-Hawking scenario our universe is produced by the quantum tunneling of a prior universe through the region of chaos. Or, as we have seen, when we properly account for the opposite arrows of time, two universes tunnel out of the chaos. In the Vilenkin scenario no prior universe exists and our line universe simply tunnels out of the chaos. In either case, a universe or universes are created *ex nihilo* by natural rather than supernatural means.

In each case we have a wave function representing the state of the universe in both the physical (something) and the nonphysical (nothing) regions. This can be used to calculate relative probability of finding a universe in an ensemble of similar universes in either the physical or the nonphysical state. The result is that the universe, for these two models, is about twice as likely to be something rather than nothing.

I do not claim that either the Hartle-Hawking or Vilenkin scenario represents exactly how the universe came to be. Each model is consistent with what we know about physics and cosmology and serves to at least show that quantum tunneling is a possible mechanism for the natural origin of our universe. Quantum tunneling is equivalent to the more vague concept of a "quantum fluctuation," which one often reads about in cosmological literature.

Whether or not a previous universe existed, we have seen that modern cosmology strongly suggests that the universe within which we reside began in a state of total disorder. Thus, any memory of any prior creation, if we can even use the word "prior" meaningfully, was lost in that chaos. A deist god who tossed the dice once and then took off to do better things may exist. But for us denizens of this universe, it has no effect and so might as well not exist.

DESCRIBING OBJECTIVE REALITY

I have demonstrated that a natural origin is possible without violating any laws of physics. So then the next question is, where did those laws come from? Didn't they have to come from God? Let's think about what the laws of physics really mean.

In this book I have implicitly referred to the laws of nature in the traditional way, as rules for the behavior of matter so that if a creator existed, the laws were mandated by that creator. Almost all theologians hold to that view. Although a majority of scientists do not believe in a creator, they still look at the laws of nature as some part of the inherent structure of the universe, governing the universe the way the US Constitution governs America.

However, philosophers of science are beginning to take another perspective on the laws of nature, which is the one to which I subscribe and proposed independently in *The Comprehensible Cosmos.* As philosopher David Armstrong has said,

> There is one truly eccentric view. This is the view that, although there are regularities in the world, there are no laws of nature.[21]

I have argued that what are called the laws of physics, which are regarded as the most basic of the laws of nature, are not restrictions on the behavior of matter but rather restrictions on what physicists can do when they invent mathematical models to describe the observed behavior of matter.

The new physics of Copernicus, Galileo, and Newton broke with the teachings of Aristotle and other ancients who had imagined two kinds of motion: (1) motion on Earth, where the "natural" movement of bodies is either up or down to the center of Earth in a straight line and (2) motion in the heavens, where the natural movement is in circles about Earth. In both cases, a preferred observer exists whose point of view is supreme—the human observer standing at rest on Earth.

Copernicus moved the preferred observer to the center of the sun. Although it took physicists centuries to grasp the full significance of the concept, Galileo's realization that motion is relative meant that there should be no preferred observer, no special *reference frame* from which we can describe observations. And, when you think about it, this makes sense. If an objective reality exists out there that is not just a fantasy or dream, then we should be able to describe that reality in a way that does not depend on a particular point of view.

If we are physicists writing down mathematical models that we expect to describe accurately an objective reality, those models must apply equally well in all frames of reference. This is especially true of any basic principles that we wish to apply universally to all or at least a wide range of different phenomena. For example, any laws of motion we propose must apply to balls, rocks, ballistic projectiles, spaceships, and monkeys jumping from trees. They must be the same on Earth, on the moon, and in the most distant galaxy.

The models and laws of physics also cannot depend on the particular moment in time we make a measurement. Again, to be objective, it should not matter whether the reference frame in which measurements are being made is that of the author of this book, Galileo, Aristotle, or some primitive caveman. They must be the same yesterday, today, tomorrow, and billions years in the past, and a billion years in the future.

Likewise, physics should not depend on the observer's orientation. It should be the same for a Canadian and an Australian.

In short, the models, theories, and laws of physics must possess what I call *point-of-view invariance.* They must be *invariant*, or unchanging, as we alter the origin, orientation, and velocity of the reference frame we use to specify position.

Invariance is closely related to the concept of *symmetry*. A sphere can be rotated about any axis through its center without changing its appearance. Thus, we say it is invariant to rotations and possesses *rotational symmetry* around all possible axes. A cylinder and cone have rotational symmetry about a single axis. I will use the terms invariance and symmetry interchangeably.

The three space-time symmetries we have discussed so far are called:

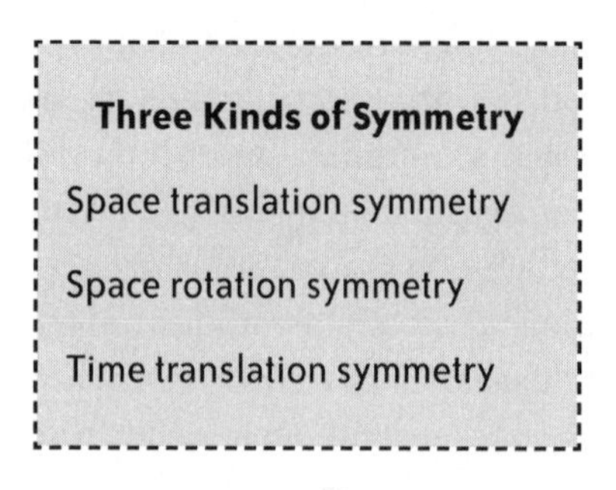

Space translation refers to moving, or "translating," the origin of a spatial coordinate system from one place in space to another. Space rotation refers to rotating the spatial coordinate axes from one orientation to another. Time translation refers to moving the point in time you define as the origin of your time axis, $t = 0$, from one moment to another. If a model is to have any one of these symmetries, it must be invariant to the corresponding transformation of the coordinate system by any arbitrary amount.

NOETHER'S THEOREM

In 1918 German mathematician Emmy Noether proved a remarkable theorem that should be more widely recognized as one of the most important discoveries of the twentieth century.[22] Noether proved that a physical model that possesses any of the three space-time symmetries I have defined will automatically contain measurable quantities that are *conserved*, that is, quantities that will not change as the system evolves with time. Specifically, Noether showed:

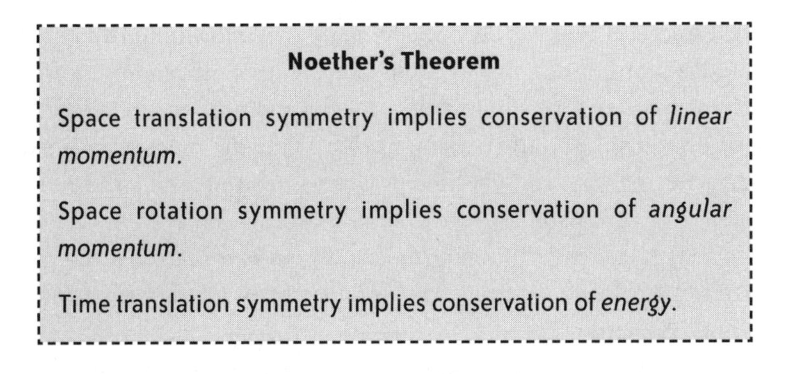

Noether's Theorem

Space translation symmetry implies conservation of *linear momentum*.

Space rotation symmetry implies conservation of *angular momentum*.

Time translation symmetry implies conservation of *energy*.

These three conservation principles, or "laws," are the most important in physics! Linear momentum is what Newton, following Descartes, called the *quantity of motion*. Examples of the three conservation principles were given in chapter 5.

THE ORIGIN OF PHYSICAL LAW

Since Newton's laws of motion (see chapter 6) can be derived from momentum conservation, classical mechanics can also be seen as a consequence of point-of-view invariance.

The revolutionary implication of Noether's theorem is that these basic laws of classical physics are not handed down from above. Nor are they, as many scientists and philosophers still believe, inherent properties of the universe that require matter to behave in a certain way. Rather, these laws are the logical consequence of physicists being required to behave a certain way. If physicists wish to formulate their models in a way that does not depend on the point of view in space and time of the observer, then those models will automatically have imbedded in them the three great conservation laws.

Newton's laws of mechanics do not specify that a body must follow this particular set of rules. Rather, the laws are statements that are required to have the form they do if they are to describe nature in a way that is independent of point of view. If they did not, then this would imply that nature itself was not point-of-view invariant, that there existed a special point of view.

For example, suppose someday it is discovered that energy is not conserved, a proposal that has been made many times to explain strange new phenomena until other explanations were found. If energy were not conserved, we could conclude that a special moment in time does exist, perhaps provided by God. You might think that the creation was such a moment. But energy conservation was not violated when the universe exploded out of chaos. We now have accurate measurement of the average energy density of the universe and we find it is zero, with the positive energy of motion and rest energy balanced by the negative potential energy of gravitation.

The fact that the conservation laws successfully describe what is observed in the universe can be taken as evidence that no special point of view exists in space-time. And, this is exactly what one would expect if there were no God or other force external to the universe establishing a particular point of view. Indeed, the laws of physics look just like they should look if the universe came from nothing. There is no special point of view in nothing.

EINSTEIN'S RELATIVITY

As we saw in chapter 8, Einstein's special theory of relativity, published in 1905, showed that spatial intervals measured with meter sticks and the time intervals measured with clocks will be different for two observers moving relative to one another. In particular, a clock that is moving with respect to an observer will appear to slow down (*time dilation*) compared to the clock in the observer's reference frame; a body that is moving with respect to an observer will appear to contract along its direction of motion (*Lorentz-Fitzgerald contraction*), as measured with a meter stick in the observer's reference frame.

These effects are tiny at low velocities, though still measurable with today's atomic clocks. They become more pronounced as the relative speeds of two observers approach the speed of light.

The relativity of space and time is supported by a century of experiments that confirm the equations of special relativity to high precision. This fact is not easily reconciled with a model of objective reality that contains space and time as substantial elements. On the other hand, relativity conforms nicely to a model of reality in which space and time are simply human contrivances, quantities measured with meter sticks and clocks. This is not to say that these concepts are arbitrary; they help describe objective observations that presumably reflect an underlying reality. However, we should not simply assume that those observations coincide with that reality—whatever it may be.

Einstein showed how to calculate space and time intervals measured in one reference frame in terms of the intervals measured in another reference frame. He used a formula derived by Hendrik Lorentz in 1892 called the *Lorentz transformation.* In 1907 Hermann Minkowski helped place special relativity on an elegant mathematical foundation by introducing the notion of four-dimensional *space-time,* where time is included as one of the four coordinates. This neatly accommodates the fact that in order to define the position of an event you need three numbers to specify its position in familiar three-dimensional space and another number to specify the event's time of occurrence. Thus, a four-dimensional vector, that is, a vector in four-dimensional space, represents the space-time position of an event. Three of the axes in this space are the conventional x, y, z Cartesian coordinates of familiar three-dimensional space. The fourth axis represents time. For geometrical convenience, the time axis is sometimes represented by ict, where $i = \sqrt{-1}$, c is the speed of light, t is the time measured on a clock. This makes the four-dimensional space Euclidean so that the square of magnitude (length) of the vector is equal to the sum of the squares of its projections on each of the four axes (four-dimensional Pythagorean theorem).

For simplicity of illustration, a two-dimensional projection of four-dimensional space-time is presented in figure 16.4, with the time axis and one space axis shown. In (a) the coordinate system (x', ict') that corresponds to a translation of the original coordinate system along the time axis is also shown. Actually, $t' = t$, and I have displaced the axes slightly for illustration. As described above, any physics model that is invariant to a time translation will automatically contain a quantity called the energy that is conserved.

In (b) a coordinate system (x', ict') is shown that corresponds to a rotation of the original coordinate system in four-dimensional space around the y, z plane. The Lorentz transformation can be shown to be equivalent to this rotation in Minkowski space. It follows that Lorentz invariance is equivalent to rotational invariance in space-time, which corresponds to a

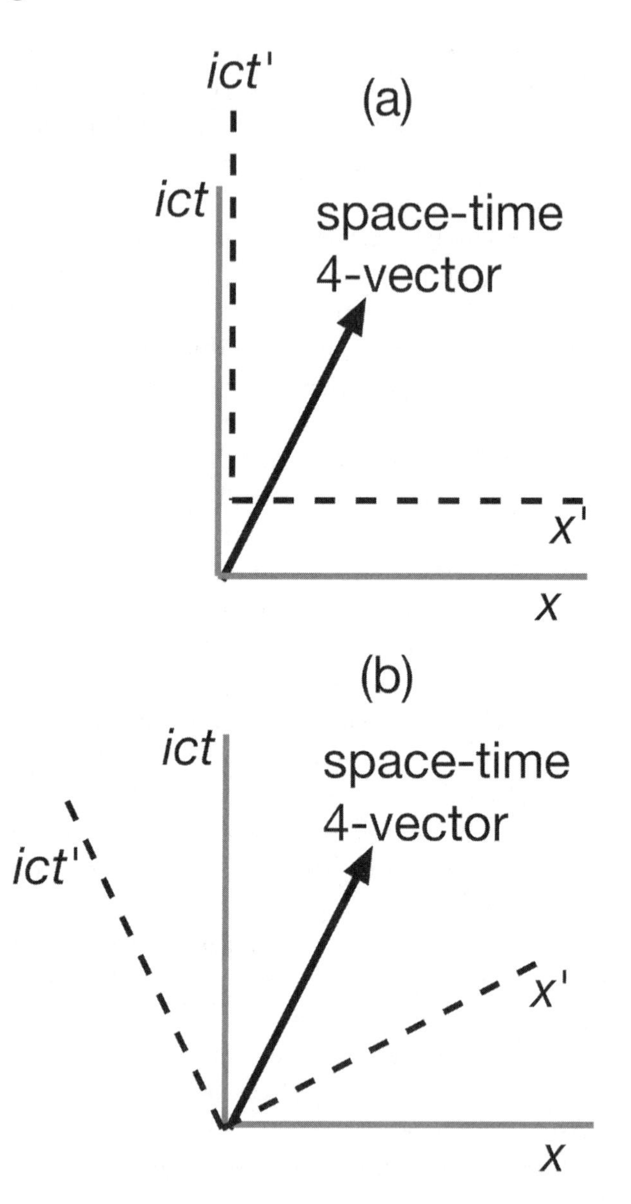

Fig. 16.4. A position vector in space-time. Only two of the four dimensions are shown. The time axis is plotted as *ict*. In (a) a translational along the time axis (slightly displaced for clarity) is shown. If the position vector does not change under this operation, energy will be automatically conserved. In (b) a rotation in the *y*, *z* plane is shown. If the position vector does not change under this operation, then the rules of special relativity will be automatically enforced.

change in coordinates while the space-time position vector remains unchanged. Any physics model that is invariant to such rotations in space-time will automatically obey all the rules of special relativity.

In short, conservation of energy, linear momentum, angular momentum, Newton's laws of motion, and all of special relativity follow from point-of-view invariance, where changing one's point of view corresponds to a translation or rotation in space-time. Galileo's relativity is also accommodated since it is subsumed in Einstein's relativity for the case where relative speeds are low compared to light.

GAUGE SYMMETRY

In my 2006 book, *The Comprehensible Cosmos: Where Do The Laws Of Physics Come From?* I showed that point-of-view invariance and Noether's theorem can be generalized to the abstract space physicists use to formulate their theories. The "state" of a physical system can be represented by a vector in this abstract space analogous to the representation of an event as a vector in space-time. A rotation of coordinate system in the abstract space is called a *gauge transformation.* When the state vector is unchanged under a gauge transformation we have *gauge invariance,* or *gauge symmetry.* When the rotation is by an amount that does not depend on space and time, it is called a *global gauge transformation.* When the rotation is different at different points in space and time, it is called a *local gauge transformation.* Virtually all of basic physics can be shown to follow from the assumption of gauge invariance—which we note is simply another form of point-of-view invariance—with few other assumptions that are mostly just matters of mathematical definition. The basic laws include conservation of electric charge, both classical and quantum electrodynamics, and the fact that the photon has zero mass. It also includes quantum mechanics, with the uncertainty principle and the superposition principle. The latter says that states can be written as linear combinations of other states and is responsible for many of the results people regard as strange, such as *entanglement.* I will not complicate matters further by getting into that.

The highly successful standard model of elementary particles and forces that was described earlier in this book is what is called a *gauge theory.* The underlying equations of the standard model are gauge invariant.

However, when it comes to applying it to laboratory observations, things are not so simple.

Gauge invariance in the standard model must be broken in order to find the equations that successfully describe actual observations. This can be understood from the point of view of point-of-view invariance. The measurements we make even with our highest-energy particle colliders represent a special "low-energy" point of view when compared to the energies that existed in the early universe. Our everyday observations are taken from the perspective of a very cold universe compared to the early big bang.

Thus the following scenario suggests itself for the source of the laws of physics. Assume the universe starts out as an empty void (see above). It possesses all the possible symmetries we can imagine and so we can model this in terms of all the laws that follow naturally from these symmetry principles. As the universe expands and cools, some of these symmetries spontaneously (that is, randomly, without cause) break and we obtain the standard model that fits all the data as of this writing.

In short, the *laws of physics are natural.* They are just what they should be if the universe appeared from a state of maximum indeterministic chaos. Now, the theist will argue that God made that chaos. But he didn't have to. As we saw previously, that chaos was natural.

NOT MAINSTREAM

I need to comment on the reception of the notion that the laws of physics follow from point-of-view invariance. Critics point out that this is not an accepted principle within mainstream physics. While this is true, it has not been rejected either. Nothing I have said conflicts with existing physics. No one has pointed to a single error in the mathematics presented in *The Comprehensible Cosmos.*[23] All I have done is give an unconventional *philosophical* interpretation to otherwise well-established theory.

Most of the great physicists of the first half of the twentieth century—Einstein, Bohr, Schrödinger, Heisenberg, and others—who brought us relativity and quantum mechanics, were very interested in the philosophical significance of their discoveries. By contrast, most of the great physicists of the second half of the century—Schwinger, Feynman, Gell-Mann, Weinberg, Glashow, and others—who brought us quantum electrodynamics and the standard model, saw little value in philosophizing. Following their lead, most contemporary physicists have adopted a "shut up and calculate" attitude in which all that a theory needs to do is produce results that agree with the data. In their view, long-winded dissertations on "what it all means" contributes nothing to this goal.

However, when the subject is not pure, unadulterated physics, but physics as applied to philosophy and theology, then interpretation is everything—after getting the physics correct.

Gauge invariance has been the guiding principle in developing the highly successful standard model of particles and forces. All I have done is give gauge invariance a more descriptive name: point-of-view-invariance, which makes its true source clear and emphasizes the connection with Noether's theorem. And, I have shown that what are called the laws of physics are the inventions of humans, not God.

SUMMARY

- The omniscient, omnipotent, omnibenevolent Judeo-Christian-Islamic God who intervenes regularly in the universe and in the lives of humans can be proved not to exist beyond a reasonable doubt. Such a God is not only logically impossible, he is falsified by the data.
- The Enlightenment deist god, who created a perfectly predetermined universe, can almost but not quite be ruled out. Both the Bohmian and many-worlds interpretations of quantum mechanics imply a deterministic universe. However, the fact that Bohm's model violates special relativity makes it unlikely to be correct. In the many-worlds interpretation, all possibilities exist and so are predetermined. The apparent randomness and free will we see in a single world is an artifact. Theology has yet to come to grips with that possibility. And science has no reason to introduce into its explanatory systems an Enlightenment deist god.
- Modern physics, including the uncertainty of conventional interpretations of quantum mechanics and deterministic chaos theory, do not provide a viable way for the Judeo-Christian-Islamic God, or any modeled after him, to intervene regularly in the universe

without noticeably breaking the laws of physics. But he could surreptitiously intervene to prevent many diseases and catastrophes, so the fact that he does not also counts against his existence.

- A new deist god consistent with statistical quantum mechanics is still possible. Almost as many Americans have a deist view of divinity as have the traditional Judeo-Christian view. The new deist god played dice with the universe and so included in its structure a high element of chance. It then left the universe alone to carry on by natural processes. Humanity has what masquerades as free will, but it is really just random—not a divine creation and so unlikely to have any divine purpose. If the universe began in total chaos, then the universe has retained no memory of any purpose of the new deist god.
- The claim that quantum mechanics shows that we can make our own reality in our minds and those minds are connected holistically to a grand unified cosmic consciousness is based on either misunderstandings or deliberate misrepresentations of what quantum mechanics really says. No empirical evidence supports the notion that mind is anything other than the product of purely material forces.
- The physics of elementary particles is not used to derive any of the principles that are observed for systems with large numbers of particles outside the realm of elementary particle physics. Condensed matter physicists, chemists, biologists, neuroscientists, sociologists, economists, and historians develop their own principles to describe their own subject matter without paying attention to particle physics. These principles are said to "emerge" from matter and "explanatory arrows" go from bottom to top. The stronger claim is that emergent principles have explanatory power going from top to bottom, thus opening up a place for God to act in the universe. No evidence for top-down causality exists and computer simulations support totally reductive, purely material bottom-up emergence.
- The laws of physics were not handed down from above but are human inventions. They take the form they do in order to guarantee that they describe observations invariant to any particular point of view. Some laws spontaneously break that symmetry, but they do so by accident.
- The model in which the universe is made of matter and nothing else

> and had a spontaneous, uncaused, natural origin from a state of chaos equivalent to "nothing" agrees with all the data. As a state of the universe, "something" is more natural than "nothing."

So we appear to have good evidence for a universe that came about spontaneously, without cause, from nothing. The laws of physics also came from nothing. The structure of the universe emerged from nothing. Indeed, we can view that structure, including Earth and humanity, as forms of frozen nothing.

NOTES

1. Robert Crease and Charles Mann, *The Second Creation* (New York: Macmillan, 1986), p. 405.

2. Victor J. Stenger, *God: The Failed Hypothesis—How Science Shows That God Does Not Exist* (Amherst, NY: Prometheus Books, 2007), ch. 8.

3. Theodore M. Drange, *Nonbelief and Evil: Two Arguments for the Nonexistence of God* (Amherst, NY: Prometheus Books, 1998).

4. John L. Schellenberg, *Divine Hiddenness and Human Reason* (Ithaca, NY: Cornell University Press, 1993).

5. Michael Martin and Ricky Monnier, eds., *The Impossibility of God* (Amherst, NY: Prometheus Books, 2003).

6. Ibid.

7. Chris Mooney, *The Republican War on Science* (New York: Basic Books, 2005).

8. David Berlinski, *The Devil's Delusion: Atheism and Its Scientific Pretensions* (New York: Crown Forum, 2008).

9. James B. Hartle and Stephen W. Hawking, "Wave Function of the Universe," *Physical Review* D28 (1983): 2960–75.

10. David Atkatz, "Quantum Cosmology for Pedestrians," *American Journal of Physics* 62 (1994): 619–27.

11. Victor J. Stenger, *The Comprehensible Cosmos: Where Do the Laws of Physics Come From?* (Amherst, NY: Prometheus Books, 2006), pp. 312–19.

12. Victor J. Stenger, "A Scenario for a Natural Origin of Our Universe," *Philo* 9, no. 2 (2006): 93–102.

13. Berlinski, *The Devil's Delusion*, p. 97.

14. The Planck mass, 2.18×10^{-8} kilogram, is the mass of a body with a radius equal to the Planck length, 1.62×10^{-35} meter.

15. Alexander Vilenkin, "Boundary Conditions and Quantum Cosmology," *Physical Review* D33 (1986): 3560–69.

16. Stenger, *The Comprehensible Cosmos*, p. 298.

17. Gottfried Wilhelm Leibniz, "Principles of Nature and of Grace Founded on Reason," in *Leibniz: Philosophical Writings*, trans. G. H. R Parkinson and M. Morris and ed. G. H. R. Parkinson (London: J. M. Dent & Sons, 1793).

18. Bede Rundle, *Why There Is Something Rather Than Nothing* (Oxford: Clarendon Press, 2004).

19. Adolf Grünbaum, "The Poverty of Theistic Cosmology," *British Journal for the Philosophy of Science* 55 (2004): 561–614, axh401.

20. Frank Wilczek, "The Cosmic Asymmetry between Matter and Antimatter," *Scientific American* 243, no. 6 (1980): 82–90.

21. David Armstrong, *What Is a Law of Nature?* (Cambridge: Cambridge University Press, 1983).

22. Nina Byers, "E. Noether's Discovery of the Deep Connection between Symmetries and Conservation Laws," *Israel Mathematical Conference Proceedings* 12 (1999), http://www.physics.ucla.edu/~cwp/articles/noether.asg/noether.html (accessed June 14, 2008). This contains links to Noether's original paper, including an English translation.

23. The first printing had a large number of typographical errors, mostly in the equations, that have been largely corrected in later printings. An erratum file can be found at http://www.colorado.edu/philosophy/vstenger/Nothing/Errata_ComprehensibleCosmos.pdf.

BIBLIOGRAPHY

Antoniadis, I., J. Ellis, J. Hagelin, and D. V. Nanopoulos. *Physics Letters* B 194 (1987) 231.

Armstrong, Karen. *The Battle for God.* New York: Ballantine Books, 2000.

Arntz, William, Betsy Chasse, and Mark Vicente. *What the Bleep!? Extended Director's Cut.* Lord of the Wind Films, LLC, 2004. www.whatthebleep.com (accessed May 12, 2008).

Aspect, Alain, Phillipe Grangier, and Roger Gerard. "Experimental Realization of the Einstein-Podolsky-Rosen *Gedankenexperiment*: A New Violation of Bell's Inequalities." *Physical Review Letters* 49 (1982): 91–94

———. "Experimental Tests of Bell's Inequalities Using Time-Varying Analyzers." *Physical Review Letters* 49 (1982): 1804–1809.

Atkatz, David. "Quantum Cosmology for Pedestrians." *American Journal of Physics* 62 (1994): 619–27.

Ayer, Alfred Jules. *Logical Positivism.* New York: Free Press, 1959.

Ball, Philip. *Critical Mass: How One Thing Leads to Another.* New York: Farrar, Straus and Giroux, 2004.

———. *The Self-Made Tapestry: Pattern Formation in Nature.* Oxford: Oxford University Press, 1999.

Barbour, Ian G. *Religion and Science.* New York: HarperCollins, 1997.

Barr, Stephen. *Physics Letters* B 112 (1982): 219.

Bartholomew, David J. *God, Chance, and Purpose: Can God Have It Both Ways?* Cambridge: Cambridge University Press, 2008.

Baylor Institute for Studies of Religion. "Baylor Religion Survey." September 2006. Selected findings at http://www.baylor.edu/content/services/document.php/33304.pdf (accessed May 12, 2008).

Bell, John S. "On the Einstein Podolsky Rosen Paradox." *Physics* 1, no. 3 (1964): 195–200.

———. *Speakable and Unspeakable in Quantum Mechanics.* Cambridge: Cambridge University Press, 1987.

Benson, Herbert M. D. *The Relaxation Response.* New York: HarperCollins, 2000. First published in 1975.

Berlinski, David. *The Devil's Delusion: Atheism and Its Scientific Pretensions.* New York: Crown Forum, 2008.

Bohm, David. *Quantum Theory.* New York: Prentice Hall, 1951.

———. "A Suggested Interpretation of Quantum Theory in Terms of 'Hidden Variables,' I and II." *Physical Review* 85 (1952): 166.

———. *Wholeness and the Implicate Order.* London: Routledge and Kegan Paul, 1980.

Bohm, David, and Basil J. Hiley. *The Undivided Universe: An Ontological Interpretations of Quantum Mechanics.* London, New York: Routledge, 1993.

Bohr, Niels. "Can the Quantum Mechanical Description of Physical Reality Be Considered Complete?" *Physical Review* 48 (1935): 696–702.

Born, Max. *The Born-Einstein Letters.* Translated by Irene Born. New York: Walker and Co., 1971.

Briggs, John, and F. David Peat. *Turbulent Mirror: An Illustrated Guide to Chaos Theory and the Science of Wholeness.* New York: Harper & Row, 1989.

Bruce, Alexandra. *Beyond the Bleep: The Definitive Unauthorized Guide to* What the Bleep Do We Know!? New York: Disinformation Company, 2005.

Byers, Nina. "E. Noether's Discovery of the Deep Connection between Symmetries and Conservation Laws." *Israel Mathematical Conference Proceedings* 12 (1999), http://www.physics.ucla.edu/~cwp/articles/noether.asg/noether.html (accessed November 5, 2004). This contains links to Noether's original paper, including an English translation.

Byrne, Rhonda. *The Secret.* New York: Attria Books, 2006.

Capra, Fritjof. *The Tao of Physics.* Boulder, CO: Shambhala, 1975.

Carnap, Rudolf. "Testability and Meaning." *Philosophy of Science* B3 (1936): 19–21; B4 (1937): 1–40.

Carpenter, W. B. "Spiritualism and Its Recent Converts." *Quarterly Review* 131: 301–53.

Chopra, Deepak. *Ageless Body, Timeless Mind: The Quantum Alternative to Growing Old.* New York: Random House, 1993.

———. *Life after Death: The Burden of Proof.* New York: Harmony Books, 2006.

———. *Quantum Healing: Exploring the Frontiers of Mind/Body Medicine.* New York: Bantam: 1989.

Chown, Marcus. *The Never-Ending Days of Being Dead: Dispatches from the Front Line of Science.* London: Faber and Faber, 2007.

Clayton, Philip. *Mind and Emergence: From Quantum to Consciousness.* Oxford: Oxford University Press, 2004.

Conway Morris, Simon. *The Crucible of Creation: The Burgess Shale and the Rise of Animals.* Oxford: Oxford University Press, 1998.

———. *Life's Solution: Inevitable Humans in a Lonely Universe.* Cambridge: Cambridge University Press, 2003.

Cooke, Bill. "Deism." In *The New Encyclopedia of Unbelief*, edited by Tom Flynn. Amherst, NY: Prometheus Books, 2007.

Craig, William Lane. "The Existence of God and the Beginning of the Universe." *Truth: A Journal of Modern Thought* 3 (1991): 85–96, http://www.leaderu.com/truth/3truth11.html (accessed July 31, 2008).

———. *The Kalām Cosmological Argument.* Library of Philosophy and Religion. New York: Macmillan, 1979.

Crease, Robert, and Charles Mann. *The Second Creation.* New York: Macmillan, 1986.

Dalai Lama. *The Universe in a Single Atom: The Convergence of Science and Spirituality.* New York: Morgan Road Books, 2005.

Davies, Paul. *The Cosmic Blueprint: New Discoveries in Nature's Creative Ability to Order the Universe.* New York: Simon & Schuster, 1988; paperback ed., Radnor, PA: Templeton Foundation Press, 2004.

———. "Teleology without Teleology: Purpose through Emergent Complexity." In *Evolution and Molecular Biology: Scientific Perspectives on Divine Action*, edited by R. J. Russell, W. R. Stoeger, F. Ayala. Vatican City: Vatican Observatory, Berkeley, CA: Center for Theology and the Natural Sciences, 1998.

Dembski, William. *Intelligent Design: The Bridge between Science and Theology.* Downer's Grove, IL: InterVarsity Press, 1999.

De Muynick, Willem M. *Foundations of Quantum Mechanics: An Empiricist Approach.* New York: Springer, 2002.

Derendinger, J., J. Kim, and D. V. Nanopoulos. *Physics Letters* B139 (1984): 170.

Deutsch, David. *The Fabric of Reality: The Science of Parallel Universes—And Its Implications.* London: Allen Lane, 1997.

DeWitt, Bryce S., and Neill Graham, eds. *The Many-Worlds Interpretation of Quantum Mechanics.* Princeton, NJ: Princeton University Press, 1973.

Dirac, Paul. *The Principles of Quantum Mechanics.* Oxford: Oxford University Press, 1930.

Donaldson, D. J., H. Tervahattu, A. F. Tuck, and V. Vaida. "Organic Aerosols and the Origin of Life: An Hypothesis." *Origins of Life and Evolution of the Biosphere* 34 (2004): 57–67.

Drange, Theodore M. *Nonbelief and Evil: Two Arguments for the Nonexistence of God.* Amherst, NY: Prometheus Books, 1998.

D'Souza, Dinesh. *What's So Great about Christianity?* Washington, DC: Regenery, 2007.

Durant, Will and Ariel. *The Story of Civilization: Part IV, The Age of Voltaire.* New York: Simon & Schuster, 1965.

Edis, Taner. *An Illusion of Harmony: Science and Religion in Islam.* Amherst, NY: Prometheus Books, 2007.

Einstein, Albert, Boris Podolsky, and Nathan Rosen. "Can the Quantum Mechanical Description of Physical Reality Be Considered Complete?" *Physical Review* 47 (1935): 777–80.

Elvee, Richard Q., ed. *Mind in Nature.* New York: Harper and Row, 1982.

Everett, Hugh, III. "Relative State Formulation of Quantum Mechanics." *Reviews of Modern Physics* 29 (1957): 454–62.

Feynman, Richard P. "The Principle of Least Action in Quantum Mechanics." PhD diss. Princeton, NJ: University Microfilms publication no. 2948, 1942.

———. *QED: The Strange Theory of Light and Matter.* Princeton, NJ: Princeton University Press, 1985. Paperback ed., 1988.

Feynman, Richard P., and A. R. Hibbs. *Quantum Mechanics and Path Integrals.* New York: McGraw-Hill, 1965.

Flynn, Tom, ed. *The New Encyclopedia of Unbelief.* Amherst, NY: Prometheus Books, 2007.

Ford, Paul Leicester, ed. *The Writings of Thomas Jefferson*, vol. 4. New York: Putnam, 1892–99.

Forrest, Barbara, and Paul R. Gross. *Creationism's Trojan Horse: The Wedge of Intelligent Design.* Oxford: Oxford University Press, 2004.

Gell-Mann, Murray, and James P. Hartle. "Time Symmetry and Asymmetry in Quantum Mechanics and Quantum Cosmology." In *Proceedings of the 1st International A. D. Sakarov Conference on Physic* and in *Proceedings of the NATO Workshop on the Physical Origin of Time Asymmetry*, edited by J. J. Haliwell, J. Perez-Mercader, and W. Zurek.

Georgi, Howard, and Sheldon Glashow. *Physical Review Letters* 32 (1974): 438.

Gilpin, Geoff. *The Maharishi Effect: A Personal Journey through the Movement That Transformed American Spirituality.* New York: Tarcher/Penguin, 2006.

Gleick, James. *Chaos: Making a New Science.* New York: Penguin, 1987.

Goswami, Amit. *The Self-Aware Universe.* New York: Penguin, 1993.

Gould, Stephen Jay. *Wonderful Life: The Burgess Shale and the Nature of History.* New York: Norton, 1989.

Gribbin, John. *Deep Simplicity: Bringing Order to Chaos and Complexity.* New York: Random House, 2004.

Griffiths, Robert J. "Consistent Histories and the Interpretation of Quantum Mechanics." *Journal of Statistical Physics* 26 (1984): 219–72.

Grünbaum, Adolf. "The Poverty of Theistic Cosmology." *British Journal for the Philosophy of Science* 55 (2004): 561–614, axh401.

Haag, James. "Between Physicalism and Mentalism: Philip Clayton on Mind and Emergence." *Zygon* 41, no. 3 (September 2006): 633–48.

Haken, Hermann. *Synergetics, an Introduction: Nonequilibrium Phase Transitions and Self-Organization in Physics, Chemistry, and Biology.* 3rd rev. enl. ed. Vienna: Springer-Verlag, 1983.

Haliwell, J. J., J. Perez-Mercader, and W. Zurek, eds. *Proceedings of the 1st International A. D. Sakarov Conference on Physics, Moscow, May 27–31, 1991* and *Proceedings of the NATO Workshop on the Physical Origin of Time Asymmetry, Mazagon, Spain, September 30–October 4, 1991.* Cambridge: Cambridge University Press, 1992.

Hameroff, S. R., A. W. Kaszniak, and A. C. Scott, eds. *Toward a Science of Consciousness—The First Tucson Discussions and Debates.* Cambridge, MA: MIT Press.

Hansel, C. E. M. *The Search for Psychic Power: ESP and Parapsychology Revisited.* Amherst, NY: Prometheus Books, 1989.

Harris Poll #80, October 21, 2006. http://www.harrisinteractive.com/harris_poll/index.asp?PID=707 (accessed March 23, 2008).

Hartle, J. B., and S. W. Hawking. "Wave Function of the Universe." *Physical Review* D28 (1983): 2960–75.

Hawking, Stephen W. *A Brief History of Time: From the Big Bang to Black Holes.* New York: Bantam, 1988.

Hawking, Stephen W., and Roger Penrose. "The Singularities of Gravitational Collapse and Cosmology." *Proceedings of the Royal Society of London*, series A, 314 (1970): 529–48.

Henry, Granville C. *Christianity and the Images of Science.* Macon, GA: Smith & Helwys, 1998.

Heriot, Drew. *The Secret (Extended Edition).* TS Production, LLC, 2006.

Hoddeson, Lillian, Laurie Brown, Michael Riordan, and Max Dresden. *The Rise of the Standard Model: Particle Physics in the 1960s and 1970s.* Cambridge: Cambridge University Press, 1997.

Humphries, Paul. "How Properties Emerge." *Philosophy of Science* 64 (1997): 1–17

Jackson, Roy. *The God of Philosophy: An Introduction to the Philosophy of Religion.* N.p.: Philosophers Magazine, 2001.

Jahn, Robert. "The Persistent Paradox of Psychic Phenomena: An Engineering Perspective." *Proceedings of the IEEE* 70 (1982): 135–70.

Jeffers, Stanley. "PEAR Lab Closes, Ending Decades of Psychic Research." *Skeptical Inquirer* (May/June 2007): 16–17.

Kauffman, Stuart A. *At Home in the Universe: The Search for the Laws of Self-Organization and Complexity.* Oxford: Oxford University Press, 1995.

———. *Reinventing the Sacred: A New View of Science, Reason, and Religion.* New York: Basic Books, 2008.

Kline, Stephen J. *Conceptual Foundations of Multidisciplinary Thinking.* Palo Alto, CA: Stanford University Press: 1995.

Knight, J. Z. "The Akashic Record and the Quantum Field." In *Parallel Lifetimes: Fluctuations in the Quantum Field.* Fireside Series, vol. 3, no. 3. N.p.: JZK Publishing, 2003–2005. http://ramtha.com/html/community/teachings/Akasha_and_Bohm.pdf (accessed May 12, 2008).

Krucoff, M. W., S. W. Crater, et al. "Music, Imagery, Touch, and Prayer as Adjuncts to Interventional Cardiac Care: The Monitoring and Actualization of Noetic Trainings (MANTRA) II Randomized Study." *Lancet* 366 (July 16, 2005): 211–17.

Kuhn, Thomas. *The Copernican Revolution: Planetary Astronomy in the Development of Western Thought.* Cambridge, MA: Harvard University Press, 1957.

Leibniz, Gottfried Wilhelm. "Principles of Nature and of Grace Founded on Reason." In *Leibniz: Philosophical Writings*, translated by G. H. R Parkinson and M. Morris and edited by G. H. R. Parkinson. London: J. M. Dent & Sons, 1793.

Madelung, Erwin. "Quantentheorie in Hydrodynamischer Form." *Zeitschrift für Physik* 43 (1927): 354–57.

Mano Singham. "The Copernican Myths." *Physics Today* (December 2007).

Martin, Michael, and Ricky Monnier, eds. *The Impossibility of God.* Amherst, NY: Prometheus Books, 2003.

———, eds. *The Improbability of God.* Amherst, NY: Prometheus Books, 2007.

Miller, Kenneth R. *Finding Darwin's God: A Scientist's Search for a Common Ground between God and Evolution.* New York: HarperCollins, 1999.

Mone, Gregory. "Cult Science: Dressing Up Mysticism as Quantum Physics." *Popular Science*, October 2004. http://www.popsci.com/scitech/article/2004-10/cult-science (accessed May 12, 2008).

Monod, Jacques. *Chance and Necessity.* Translated by Austryn Wainhouse. New York: Collins, 1972.

Mooney, Chris. *The Republican War on Science.* New York: Basic Books, 2005.

National Academy of Sciences. *Teaching about Evolution and the Nature of Science.* Washington, DC: National Academy of Sciences, 1998, http://www.nap.edu/catalog/5787.html (accessed March 5, 2006).

O'Murchu, Diarmuid. *Quantum Theology: Spiritual Implications of the New Physics.* New York: Crossroad, 1997.

Oppy, Graham. Review of *A Case against Accident and Self-Organization* by Dean L. Overman. *Secular Web* (1999), http://www.infidels.org/zlibrary/modern/graham_oppy/overman.html (accessed May 14, 2008).

Overman, Dean L. *A Case against Accident and Self-Organization.* New York: Rowman & Littlefield, 1997.

Overstreet, A. T. "How a False Doctrine Could Persist So Long." In *The Gospel Truth: Are Men Born Sinners?* ch. 6, http://www.gospeltruth.net/menbornsinners/mbs06.htm (accessed May 12, 2008).

Paley, William. *Natural Theology or Evidences of the Existence and Attributes of the Deity Collected from the Appearance of Nature.* London: Halliwell, 1802.

Penrose, Roger. *The Emperor's New Mind: Concerning Computers, Minds, and the Laws of Physics.* Oxford: Oxford University Press, 1989.

———. *Shadows of the Mind: A Search for the Missing Science of Consciousness.* Oxford: Oxford University Press, 1994.

Penrose, Roger, and Stuart R. Hameroff. "Orchestrated Objective Reduction of Quantum Coherence in Brain Microtubules: The 'ORCH OR' Model for Consciousness." In *Toward a Science of Consciousness*, edited by Stuart R. Hameroff, Alfred W. Kaszniak, Alwyn C. Scott. Cambridge, MA: MIT Press, 1996.

Peters, Ted, and Nathan Hallanger, eds. *God's Action in Nature's World: Essays in Honour of Robert John Russell.* Williston, VT: Ashgate, 2006.

Pew Global Attitudes Project. "World Publics Welcome Global Trade—But Not Immigration." Pew Research Center, October 4, 2007, http://pewglobal.org/reports/pdf/258.pdf (accessed May 12, 2008).

Polanyi, Michael. "Life's Irreducible Structure." In *Knowing and Being: Essays by Michael Polanyi*, edited by Marjorie Grene. London: Routledge and Kegan Paul, 1969.

———. *The Logic of Personal Knowledge.* New York: Free Press, 1961.

Polkinghorne, John. *Belief in God in the Age of Science.* New Haven, CT: Yale University Press, 1998.

———. "The Metaphysics of Divine Action." In *Chaos and Complexity: Scientific Perspectives on Divine Action.* Vatican City: Vatican Observatory Publications; Berkeley, CA : Center for Theology and the Natural Sciences, 1997.

———. *Quantum Physics and Theology: An Unexpected Kinship.* New Haven, CT: Yale University Press, 2007.

Pollard, William G. *Chance and Providence: God's Actions in a World Governed by Scientific Law.* London: Faber and Faber, 1958.

Pope John Paul II. "Address to the Academy of Sciences, October 28, 1986." *L'Osservatore Romano*, English ed., November 24, 1986, p. 22.

Popper, Karl. *The Logic of Scientific Discovery.* English ed. New York: Basic Books, 1959.

Poundstone, William. *The Recursive Universe.* New York: Morrow, 1985.

Prigogine, Ilya, and Isabella Stengers. *Order out of Chaos.* New York: Bantam, 1984.

Proctor, Bob. *You Were Born Rich.* LifeSuccess Products, 1997, http://www.synergylifesuccess.com (accessed July 12, 2008).

Putoff, Harold E., and Russell Targ. "A Perceptual Channel for Information Transfer over Kilometer Distances: Historical Perspective and Recent Research." *Proceedings of the IEEE* 64 (1976): 329–54.

Quantum Depth Healing. http://elevatedtherapy.org.uk/index-page12.html (accessed May 12, 2008).

Quantum-Touch. http://www.quantumtouch.com (accessed May 12, 2008).

Radin, Dean. *The Conscious Universe: The Scientific Truth of Psychic Phenomena.* New York: HarperCollins, 1997.

Robinson, Richard. "Jump Starting a Cellular World: Investigating the Origin of Life, from Soup to Networks." *Public Library of Science Biology* 3, no. 11 (November 2004): e396, http://biology.plosjournals.org/perlserv/?request=get-document&doi=10.1371%2Fjournal.pbio.0030396 (accessed June 30, 2008).

Rosen, Edward. *Copernicus and His Successors.* London: Hambledon Press, 1996.

Rundle, Bede. *Why There Is Something Rather Than Nothing.* Oxford: Clarendon Press, 2004.

Russell, Robert John, Philip Clayton, Kirk Wegter-McNelly, and John Polkinghorne, eds. *Quantum Mechanics: Scientific Perspectives on Divine Action.* Vatican City: Vatican Observatory Publications, 2001.

Russell, Robert John, Nancey Murphy, and Arthur Peacocke, eds. *Chaos and Complexity: Scientific Perspectives on Divine Action.* Vatican City: Vatican Observatory Publications, 1996.

Russell, Robert John, Nancey Murphy, Theo C. Meyering, and Michael A. Arbib, eds. *Neuroscience and the Person: Scientific Perspectives on Divine Action.* Vatican City: Vatican Observatory Publications, 1999.

Russell, Robert John, Nancey Murphy, Arthur Peacocke, and C. J. Isham, eds. *Quantum Cosmology and the Laws of Nature: Scientific Perspectives on Divine Action.* Vatican City: Vatican Observatory Publications, 1993.

Russell, Robert John, and William R. Stoeger, eds. *Evolutionary and Molecular Biology: Scientific Perspectives on Divine Action.* Vatican City: Vatican Observatory Publications, 1998.

Sansbury, Timothy. "The False Promise of Quantum Mechanics." *Zygon* 42, no. 1 (March 2007): 111–12.

Satinover, Jeffrey. *The Quantum Brain: The Search for Freedom and the Next Generation of Man.* New York: Wiley & Sons, 2001.

Saunders, Nicholas. *Divine Action and Modern Science.* Cambridge: Cambridge University Press, 2002.

Scheiber, Bela, and Carla Selby, eds. *Therapeutic Touch.* Amherst, NY: Prometheus Books, 2000.

Schellenberg, John L. *Divine Hiddenness and Human Reason.* Ithaca, NY: Cornell University Press, 1993.

Schroeder, Gerald L. *The Science of God: The Convergence of Scientific and Biblical Wisdom.* New York: Free Press, 1997.

Schweber, Sylvan S. *QED and the Men Who Made It.* Princeton, NJ: Princeton University Press, 1976.

Shannon, C. E. "A Mathematical Theory of Communication." *Bell System Technical Journal* 27 (July 1948): 379–423; (October 1948): 623–25.

Shannon, Claude, and Warren Weaver. *The Mathematical Theory of Communication.* Champaign: University of Illinois Press, 1949.

Shermer, Michael. "Quantum Quackery." *Scientific American* 292, no. 1 (January 2005): 234.

Smolin, Lee. *Three Roads to Quantum Gravity.* New York: Basic Books, 2001.

Sobel, Dava. *Longitude: The True Story of a Lone Genius Who Solved the Greatest Scientific Problem of His Time.* New York: Walker, 1995.

Sober, Elliot. "It Had to Happen." *New York Times*, November 30, 2003.

Sproul, R. C. *Not a Chance: The Myth of Chance in Modern Science.* Grand Rapids, MI: Baker Books, 1994.

Stapp, Henry P. "S-Matrix Interpretation of Quantum Theory." *Physical Review* D3 (March 15, 1971): 1319.

Stenger, Victor J. "Bioenergetic Fields." *Scientific Review of Alternative Medicine* 3, no. 1 (Spring/Summer 1999).

———. *The Comprehensible Cosmos: Where Do the Laws of Physics Come From?* Amherst, NY: Prometheus Books, 2007.

———. "Fitting the Bible to the Data." *Skeptical Inquirer* 23, no. 4 (1999): 67–68.

———. *God: The Failed Hypothesis—How Science Shows That God Does Not Exist.* Amherst, NY: Prometheus Books, 2007.

———. *Has Science Found God? The Latest Results in the Search for Purpose in the Universe.* Amherst, NY: Prometheus Books, 2003.

———. "In the Name of the Omega Point Singularity." *Free Inquiry* 27, no. 5 (August/September 2007): 62.

———. "Neutrino Oscillations in DUMAND." *Proceedings of the Neutrino Mass Miniconference, Telemark Wisconsin*, October 2–4, 1980, Madison, WI, University of Wisconsin Report 186, 1980.

———. *Physics and Psychics: The Search for a World Beyond the Senses.* Amherst, NY: Prometheus Books, 1990.

———. "The Premise Keepers." *Free Inquiry* 23, no. 3 (Summer 2003).

———. "A Scenario for a Natural Origin of Our Universe." *Philo* 9, no. 2 (2006): 93–102.

———. *Timeless Reality: Symmetry, Simplicity, and Multiple Universes.* Amherst, NY: Prometheus Books, 2000.

———. *The Unconscious Quantum: Metaphysics in Modern Physics and Cosmology* Amherst, NY: Prometheus Books, 1995.

Stokes, Douglas M. "The Shrinking Filedrawer: On the Validity of Statistical Meta-Analysis in Parapsychology." *Skeptical Inquirer* 25, no. 3 (2001): 22–25.

Super-Kamiokande Collaboration. "Search for Proton Decay via $p \rightarrow e^{+} \pi^{0}$ in a Large Water Cherenkov Detector." *Physical Review Letters* 81 (1998): 3319.

Susskind, Leonard. *The Cosmic Landscape: String Theory and the Illusion of Intelligent Design.* New York: Little, Brown, 2006.

Tegmark, Max. "The Importance of Quantum Decoherence in Brain Processes." *Physical Review* E61 (1999): 4194–206.

Teilhard de Chardin, Pierre. *The Phenomenon of Man.* English translation by Bernard Wall. London: Wm. Collins Sons; New York: Harper & Row, 1959. Originally published in French as *Le Phénomene Humain.* Paris: Editions du Seul, 1955.

Tipler, Frank J. *The Physics of Christianity.* New York: Doubleday, 2007.

———. *The Physics of Immortality: Modern Cosmology and the Resurrection of the Dead.* New York: Doubleday, 1994.

Tracy, Thomas F. "Divine Action and Quantum Theory." *Zygon* 35, no. 4 (December 2000): 891–900.

US Religious Landscape Survey. Pew Forum in Religion and Public Life. Pew Research Center, 2008, http://religions.pewforum.org/reports (accessed May 12, 2008).

Van Huyssteen, J. Wentzel. "Emergence and Human Uniqueness: Limiting or Delimiting Evolutionary Expmation?" *Zygon* 41, no. 3 (September 2006): 649–64.

Vilenkin, A. "Boundary Conditions and Quantum Cosmology." *Physical Review* D33 (1986): 3560–69.

Vitzthum, Richard C. *Materialism: An Affirmative History and Definition.* Amherst, NY: Prometheus Books, 1995.

Ward, Keith. *Divine Action.* London: Collins Religious Publishing, 1990.

Weinberg, Steven. *Dreams of a Final Theory: The Search for the Fundamental Laws of Nature.* New York: Random House, 1992.

Wigner, Eugene P. "The Probability of the Existence of a Self-Reproducing Unit." In Michael Polanyi, *The Logic of Personal Knowledge.* New York: Free Press, 1961.

Wilczek, Frank. "The Cosmic Asymmetry between Matter and Antimatter." *Scientific American* 243, no. 6 (1980): 82–90.

Wildman, Wesley J. "The Divine Action Project, 1988–2003." *Theology and Science* 2, no. 1 (2004): 31–75.

———. "Further Reflections on the Divine Action Project." *Theology and Science* 3, no. 1 (2005): 71–83.

Wittgenstein, Ludwig. *Philosophical Investigations.* New York: Macmillan, 1953.

———. *Tractatus Logico-Philosophicus.* Translated by C. K. Ogden. Routledge and Kegan Paul, 1922.

Wolf, Fred Alan. *Taking the Quantum Leap.* New York: HarperCollins, 1981.

———. *The Yoga of Time Travel: How the Mind Can Defeat Time.* Wheaton, IL: Quest Books, 2004.

Wolfram, Stephen. *A New Kind of Science.* Champaign, IL: Wolfram Media, 2002.

Zukav, Gary. *The Dancing Wu Li Masters: An Overview of the New Physics.* New York: Morrow, 1979.

———. *The Seat of the Soul.* New York: Fireside, 1989.

Zurek, Wojciech H. "Decoherence, Einselection, and the Quantum Origins of the Classical." *Reviews of Modern Physics* 75 (2003): 715.

INDEX

ABOUT THE AUTHOR

Victor J. Stenger grew up in a Catholic working-class neighborhood in Bayonne, New Jersey. His father was a Lithuanian immigrant, his mother the daughter of Hungarian immigrants. He attended public schools and received a bachelor's of science degree in electrical engineering from Newark College of Engineering (now New Jersey Institute of Technology) in 1956. While at NCE, he was editor of the student newspaper and received several journalism awards.

Moving to Los Angeles on a Hughes Aircraft Company fellowship, Dr. Stenger received a master's of science degree in physics from UCLA in 1959 and a PhD in physics in 1963. He then took a position on the faculty at the University of Hawaii, then retired in Colorado in 2000. He currently is emeritus professor of physics and astronomy at the University of Hawaii and adjunct professor of philosophy at the University of Colorado. Dr. Stenger is a fellow of the Committee for Skeptical Inquiry and a research fellow of the Center for Inquiry. Dr. Stenger has also held visiting positions on the faculties of the University of Heidelberg in Germany, Oxford in England (twice), and has been a visiting researcher at Rutherford Laboratory in England, the National Nuclear Physics Laboratory in Frascati, Italy, and the University of Florence in Italy.

His research career spanned the period of great progress in elementary particle physics that ultimately led to the current *standard model.* He participated in experiments that helped establish the properties of strange particles, quarks, gluons, and neutrinos. He also helped pioneer the emerging fields of very high-energy gamma-ray and neutrino astronomy. In his last project before retiring, Dr. Stenger collaborated on the underground experiment in Japan that in 1998 showed for the first time that the neutrino has mass. The Japanese leader of this experiment shared the 2002 Nobel Prize for this work.

Victor Stenger has had a parallel career as an author of critically well-received popular-level books that interface between physics and cosmology and philosophy, religion, and pseudoscience. These include: *Not by Design: The Origin of the Universe* (1988); *Physics and Psychics: The Search for a World beyond the Senses* (1990); *The Unconscious Quantum: Metaphysics in Modern Physics and Cosmology* (1995); *Timeless Reality: Symmetry, Simplicity, and Multiple Universes* (2000); *Has Science Found God? The Latest Results in the Search for Purpose in the Universe* (2003); *The Comprehensible Cosmos: Where Do the Laws of Physics Come From?* (2006); and *God: The Failed Hypothesis—How Science Shows That God Does Not Exist* (2007). The last made the *New York Times* best seller list in March 2007.

Dr. Stenger and his wife, Phylliss, have been happily married since 1962 and have two children and four grandchildren. They now live in Lafayette, Colorado. They attribute their long lives to the response of evolution to the human need for babysitters, a task they joyfully perform. Phylliss and Vic are avid doubles tennis players, generally enjoy the outdoor life in Colorado, and travel the world as often as they can.

Dr. Stenger maintains a popular Web site (a thousand hits per month), where much of his writing can be found, at http://www.colorado.edu/philosophy/vstenger/.